M000045838

# STEWARD OF THE LAND

# THE HILL COLLECTION

*Holdings of the LSU Libraries*

# STEWARD

~~~~ OF THE ~~~~

# LAND

## SELECTED WRITINGS

—— OF ——

### NINETEENTH-CENTURY
# HORTICULTURIST

# THOMAS AFFLECK

EDITED AND ANNOTATED BY
# LAKE DOUGLAS

*Louisiana State University Press    Baton Rouge*

Publication of this book was supported by a David R. Coffin Publication Grant
from the Foundation for Landscape Studies.


Published by Louisiana State University Press
Copyright © 2014 by Louisiana State University Press
All rights reserved
Manufactured in the United States of America
First printing

DESIGNER: Michelle A. Neustrom
TYPEFACE: Ingeborg
PRINTER AND BINDER: Maple Press

FRONTISPIECE PHOTO:
Thomas Affleck in 1866 or 1867. *Cushing Memorial Library and Archives, Texas A&M University.*

LIBRARY OF CONGRESS CATALOGING-IN-PUBLICATION DATA

Affleck, Thomas, 1812–1868.
Steward of the land : selected writings of nineteenth-century horticulturist Thomas Affleck / edited and annotated by Lake Douglas.
pages cm
Includes bibliographical references and index.
ISBN 978-0-8071-5810-4 (cloth : alk. paper) — ISBN 978-0-8071-5811-1 (pdf) — ISBN 978-0-8071-5812-8 (epub) — ISBN 978-0-8071-5813-5 (mobi) 1. Affleck, Thomas, 1812–1868. 2. Horticulturists—Mississippi—Biography. 3. Horticulture—United States. 4. Gardening. I. Douglas, Lake, 1949- II. Title.
SB63.A25A34 2014
635.092—dc23
[B]

2014018041

"Thomas Affleck: Missionary to the Planter, the Farmer and the Gardener,"
by Robert Webb Williams Jr., appeared first in *Agricultural History* 31, № 3 (July 1957),
and is reprinted here with permission of the editor.

Sections of Fred Carrington Cole's dissertation, "The Texas Career of Thomas Affleck"
(Louisiana State University, 1942), are reprinted here by permission of Robert Grey Cole.

The paper in this book meets the guidelines for permanence and durability of the Committee
on Production Guidelines for Book Longevity of the Council on Library Resources. ∞

—————◆—————

*Inspired by the scholarship of John Brinckerhoff Jackson*
*(1909–1996)*

*Dedicated to the scholarship of Robert Webb Williams Jr. (1923–1993)*
*and Fred Carrington Cole (1912–1986)*

—————◆—————

# CONTENTS

CHRONOLOGY
Thomas Affleck's Life and Related People and Events   245

A NOTE ON THE THOMAS AFFLECK PAPERS
*by Tara Zachary Laver*   251

INDEX   253

# INTRODUCTION

He was not only a good, but in our judgment, a great man, and at least fifty years in advance of the age in which he lived. As a man of science, he occupied a proud position, but not one equal to the greatness of his conceptions. He did much as a writer for the orchard, the garden, and all the branches of agriculture, and the record he has made of his greatness will be better appreciated now that he has left us. Because he was ahead of his age, many thought him visionary, but the day will come, when men will say that a monument should be erected to his memory.

—Thomas Affleck's obituary
*Dallas Herald,* January 16, 1869

Thomas Affleck was born in Dumfries, Scotland, in 1812, immigrated to America in 1832, and died in Brenham, Texas, in 1868. He was largely self-taught, and his interests encompassed a broad spectrum of the physical and natural sciences. While his talents as a journalist, entrepreneur, and "friend" to all brought him recognition during his lifetime and facilitated his efforts to improve regional agricultural endeavors, today he and his accomplishments remain relatively unknown.[1]

Affleck's arrival and early life in America coincided with America's physical expansion, advances in agriculture and technology, and cycles of economic boom and bust. Encouraged by a national sense of optimism, increasingly efficient means of production and transportation, and new ways to market goods and services, regional markets grew as populations moved westward from the Atlantic seaboard states into the interior territories.

1. A biographical entry on Affleck appears in Charles A. Birnbaum and Stephanie S. Foell, eds., *Shaping the American Landscape* (Charlottesville: University of Virginia Press, 2009), 5–7. For biographical summaries of Affleck's Texas years, see Samuel Wood Geiser, *Horticulture and Horticulturists in Early Texas* (Dallas: Southern Methodist University Press, 1945), 31–32, together with the Texas Historical Association's on-line resource, http://www.tshaonline.org/hand-book/online/articles/faf03, accessed July 25, 2012.

This national climate was ideal for enterprising, ambitious entrepreneurs such as young Affleck. By 1837, he and his family had settled in Clinton, Indiana, but by 1838, defeated by the deaths of his wife and child and discouraged by business failures and his own illness, Affleck moved to Cincinnati in preparation for a return to Scotland. He recuperated, however, and took a job as a junior editor for the *Western Farmer,* an agricultural paper published in Cincinnati and aimed at growing markets in the West and South. The publication included not only contributions from its subscribers, as was common at the time among such publications, but also original observations and first-hand reporting from its staff. With Affleck's agreeable and engaging writing style, the newspaper's circulation and influence grew. Soon the periodical, now known as the *Western Farmer and Gardener,* regularly featured articles on new farming techniques, time-saving agricultural implements,

Certificate from the Agricultural, Horticultural, and Botanical Society of Jefferson College in Washington, Mississippi, awarded to Anna Smith (Affleck's future wife) in 1841 for "best peaches exhibited." *Cushing Memorial Library and Archives, Texas A&M University.*

Ingleside, the Affleck home in Washington, Mississippi. Photo taken ca. 1951. *Cushing Memorial Library and Archives, Texas A&M University.*

domestic gardening, and animal husbandry, written by Affleck and a growing regional network of correspondents.

In the early 1840s, Affleck traveled the Mississippi River Valley attending agricultural fairs, meeting people, and reporting on the people, places, and trends he encountered. In Natchez, he met the widow Anna Dunbar Smith, and in 1842 he returned to Washington, a rural community outside Natchez, married Anna, and began managing her plantations. It was here that Affleck created Southern Nurseries, one of the South's first commercial nurseries.[2] He imported plants from European and American sources, acclimated them to local soil and weather conditions, and marketed them throughout the re-

2. In the 1840s, the few commercial nurseries in America were on the East Coast, although as early as the 1820s ornamental plants were advertised in New Orleans newspapers and sold in hardware stores and commercial outlets that supplied rural areas. Outside urban areas, little documentation has surfaced on regional sources for ornamental plants in the South. Itinerant peddlers and plantsmen traveled through rural areas selling plants, but, as Affleck often noted, supplies were of poor quality and largely unsuited for the locations in which they were sold.

gion. He also began an active correspondence with clients, suppliers, horticultural experts, publishers, and merchants.

Much of Affleck's commercial business involved clients and suppliers in New Orleans, which was easily accessible to Natchez by regular steamboat connections on the Mississippi River. In 1846 he became the editor of *Norman's Southern Agricultural Almanac,* which was published in New Orleans, and by 1851 he had taken over its publication. For the next decade, it appeared as *Affleck's Southern Rural Almanac and Plantation and Garden Calendar.* The almanac offered practical advice on how, when, and what to plant; reported on regional environmental conditions; and listed plant varieties available from his Southern Nurseries in Washington. These almanacs represent the largest single collection of Affleck's writings. Examples are accessible in several archives: a well-worn copy of *Norman's Southern Agricultural Almanac for 1847* and a bound volume of four *Southern Rural Almanacs* for 1851–54 are in the holdings of The Historic New Orleans Collection, and the Affleck archives at both Louisiana State University and Texas A&M University have assorted copies up to 1861. A limited edition reprint of the 1860 *Southern Rural Almanac* appeared in 1986 from the New Year's Creek Settlers Association of Brenham, Texas, and the *Almanac* for 1854, from Harvard University's Arnold Arboretum collection, is available electronically.[3] Since the aim of this book is to bring forward less accessible examples of Affleck's writings, selections from these almanacs are not included here.

In the mid-1850s Affleck established a nursery in Brenham, Texas, and by the late 1850s he had moved to Texas. Here, following the Civil War, he turned his attention to rebuilding Texas livestock herds and rejuvenating agricultural labor forces depleted by the devastation of the Civil War, activities with which he was engaged at the time of his unexpected death in late 1868. Efforts in these areas, like his earlier commercial ventures, were not financially successful; economic stability remained elusive throughout his life and continued to be a problem for his family after his death.

In addition to prolific correspondence, Affleck wrote authoritatively on numerous subjects for a variety of publications, including local and regional newspapers, agricultural journals, and books. While personal and business correspondence has formed the basis for previous academic examinations

---

3. http://books.google.com/books/reader?id=BUCAAAAYAAJ&printsec=frontcover&output =reader&pg=GBS.PP5

of Affleck's life and career (and the biographical essays reprinted in this book), Affleck's published works, for the most part, have not been examined. Representative selections are given here. My aim in gathering and annotating them is threefold: first, to enable us in the twenty-first century to gain a better understanding of agricultural and horticultural life in the mid-nineteenth-century American South through access to previously unexamined first-hand accounts; second, to establish Affleck's rightful place, alongside contemporaries such as Andrew Jackson Downing and Frederick Law Olmsted, at the beginning of America's understanding of its landscape and his role in developing a regional comprehension of the environment in general and of ornamental horticulture in particular; and finally, to present evidence that affirms the assessment of Affleck, noted in his obituary, as a "man of science" and an important "writer for the orchard, the garden, and all the branches of agriculture."

Affleck's numerous interests primarily involved the land and its stewardship. Animal husbandry, agriculture and farm management, ornamental horticulture, plant and soil science, water management, climate, and entomology are all subjects about which he wrote with the intelligence and authority of experience. Later interests, of European workers and labor represented in his post–Civil War Texas years, included activities related to immigration, manufacturing, dredging and shipping, rejuvenation of cattle stock, and food preservation. Here, his work is expressed less in published articles and more in his correspondence and newspaper accounts of the period. Affleck had an enterprising and entrepreneurial spirit (often to his detriment), a talent for written and spoken communications, and the capacity to establish personal relationships with those he met, those with whom he corresponded, and those who read (and bought) his publications. His ideas were shaped by a lifelong quest for knowledge, and his recommendations were informed by investigations as well as by the application of information gained through personal observation and the experiences of others. Ever helpful, Affleck eagerly and tirelessly corresponded with others, and through his correspondence and business archive, we find confirmation of what his writings demonstrate.

Students of American landscape architecture and garden history will see interesting and unexpected similarities in the lives and careers of Affleck (1812–68), Andrew Jackson Downing (1815–52), and Frederick Law Olmsted (1822–1903). While these intersections may be tenuous, they are worth noting:

- All three were involved in writing about the environment at multiple scales, and the careers of all three advanced because of the publication and distribution of their written works.
- Downing and Olmsted knew each other, but we do not know whether these two East Coast residents knew of Affleck and his work in the South.
- We know from Olmsted's writings that he visited Natchez and its region at the same time Affleck was living there; neither, however, describes having met the other. Perhaps Olmsted and Downing had read some of Affleck's articles; it is clear that Affleck knew of Downing and his writings.[4]
- Downing and Affleck wrote about similar domestic-oriented subjects, and each had an interest and experience in commercial nurseries and ornamental horticulture.
- Tragic boat accidents figure prominently in the lives of both Downing and Affleck: Downing's life ended at an early age (thirty-seven), well before he had reached his potential, when the steamboat on which he was traveling exploded and burned on the Hudson River; he died rescuing other passengers. Affleck died at age fifty-six, a relatively young man, discouraged perhaps but not broken by a series of devastating events in his life, one of which was the loss of his entire nursery stock in 1861 on the Mississippi River in transit from Mississippi to Texas when the steamboat on which he and his inventory were traveling burned "to the water's edge" on the Mississippi north of New Orleans. "L'homme propose, mais Dieu dispose," he noted wryly.[5]
- Unlike Downing, whose father's business became his career, neither Olmsted nor Affleck had a mentor, and both embarked in numerous career directions before settling into the ones for which they are now known. They both often met with failure; nevertheless, both started over, reinventing their professional careers and private lives. Recall that Olmsted was a merchant seaman, a "scientific" farmer, a journalist, and a park manager prior to winning the design competition, with Calvert Vaux, for

4. Downing's published work was often discussed in agricultural papers in the 1840s and 1850s; in 1853 Olmsted contributed to the *American Agriculturist*, a paper that earlier had published articles by Affleck. An advertisement for books in *Norman's Southern Agricultural Almanac for 1847*, which Affleck edited, lists four of Downing's books for sale in New Orleans by the almanac's publisher, B. F. Norman, and the *Southern Rural Almanac for 1860* has a full-page advertisement for *The Horticulturist and Journal of Rural Art*, established in 1846 by Downing.

5. Roughly translated, "Man proposes, God decides."

what became New York's Central Park. Later, he was involved in public health issues and land management before turning full time to landscape design work. Affleck's career, too, took different turns when, following defeat, he started over: first, after the tragedy of his family's death and his own bankruptcy in the late 1830s when he retreated to Cincinnati and became a journalist; second, in rural Mississippi, after his move there in 1842 to start a commercial nursery; third, when he moved his business to Texas in the mid-1850s; and finally, in the mid-1860s, when he was attempting to recover from the Civil War and devoted time to labor, live-stock, and economic-development initiatives in Texas.

- Early in their careers, both Affleck and Olmsted traveled in areas of the American West and Southwest with which they were unfamiliar, writing about what they encountered for newspaper readers, and much of what they observed concerns the landscape.

- Affleck and Olmsted shared an interest in public health and hygiene: Affleck, through the article given here ("On the Hygiene of Cotton Plantations and the Management of Negro Slaves" of 1851), and Olmsted, through his work during the Civil War with the United States Sanitary Commission.

- All three were involved in landscapes for government buildings: Affleck provided a planting design for Louisiana's capitol in Baton Rouge (1859), although the plans have not surfaced, and there is a passing reference to his having supplied plants for the Texas state capitol in Austin.[6] Down-

---

6. The New Orleans *Daily Crescent,* October 2, 1858, reported that "the work of ornamenting the grounds has been commenced; the hill is being terraced and is to be sodded." Citing this article and a later one from the Baton Rouge *Weekly Gazette & Comet* (March 25, 1864), architectural historian Arthur Scully Jr. claims that Affleck "planted trees, shrubs, and flowers of all varieties. A fountain and pool were placed at the lower end of the terrace near the river, and seats and walkways created an inviting atmosphere." See *James Dakin, Architect* (Baton Rouge: Louisiana State University Press, 1973), 151. From the New Orleans *Daily Picayune,* February 17, 1859: "Our old acquaintance, Mr. Thomas Affleck, paid us a visit yesterday. He has been engaged, we understand, to lay out the grounds around the capitol at Baton Rouge, and the selection could not have been better made. Mr. Affleck's taste and long experience in ornamental gardening fit him well for such a task." From the Baton Rouge *Weekly Gazette & Comet,* July 31, 1859: "The Grounds about the Capitol.—The work of terracing is done, and all the trees, shrubs and flowers, are growing finely. The walks have been protected with shells, and the place has the appearance of a flower garden. The neatness and order of the grounds attract the ladies and children there in the evening, and it is getting to be a fashionable place of resort." See also Carol K. Hasse, *Louisiana's Old State Capitol* (Gretna: Pelican Publishing, 2009), 26–27. Hasse does not assign credit for the design of the state capitol's grounds.

ing was engaged by President Millard Fillmore to design the grounds of the White House, Capitol, and Smithsonian Institute (1851), but he died before they were executed; and Olmsted later completed plans for the U.S. Capitol grounds and terraces (1874) and the Connecticut State House (1878), which were both installed.

- All three left extensive writings in different formats and published in different venues (including books, journal articles, journals, reports, correspondence, and business records), allowing opportunities for today's readers to consider and reflect on their words with the benefit of hindsight and comparison.

- In all three men, we see a capacity to capitalize on strengths, focus energy, re-imagine careers and re-invent personal narratives, and to embark on new ventures with enthusiasm, dedication, and optimism. These personal qualities characterize many men in mid-nineteenth-century America.

- All three must have shared personality traits that enabled them to become acquainted with elected officials and the social, economic, and intellectual leaders of their times. Each met people easily and moved effectively in various social, economic, and political circles and subsequently succeeded in using contacts among prominent decision-makers, businessmen, and influential community leaders to their benefit.

- Finally, of Affleck, Downing, and Olmsted, only the latter lived long enough to have been involved in numerous and influential projects related specifically to what we know now as landscape architecture. We may speculate that, had the others lived longer, they also might have attained professional influence in their respective areas of expertise commensurate with Olmsted.

One reason we know less about Affleck and more about Downing and Olmsted is because of the association of Downing's and Olmsted's careers with East Coast cities—New York, Boston, and Washington, D.C.—and the tendency history has of favoring events and people from those urban centers over those in the rural areas of the West (Kentucky, Ohio, Indiana, and Illinois) and Southwest (Mississippi, Louisiana, Texas, and Arkansas), where Affleck lived and worked. Until recently, this has certainly been the case in the history of American landscape architecture: for the most part, standard twentieth-century accounts of our country's landscape history, such as Norman T. Newton's iconic *Design on the Land,*[7] paid little attention to projects, events, and people outside of those on the East Coast.

More recent landscape histories are beginning to notice the people outside Olmsted and his circle of male colleagues in Boston and New York. We are now recognizing the contributions of others from the profession's early years—notably women—and beginning to understand that what is now the profession of landscape architecture was built not only by iconic figures and large projects but also by the smaller works of lesser-known people, many of whom were horticulturists, journalists, plant suppliers, and gardeners.[8]

One reason Affleck has attracted little scholarly attention, except as the subject of an occasional conference presentation or a reference in a publication, could relate to the diversity of his interests. Here comparisons with Downing's and Olmsted's reputations are again useful. For many years, Downing's importance was mainly as a horticultural writer, and only recent reappraisals of his career have repositioned him as a central feature in the rise of the American middle class and as a major influence in the development of the modern profession of landscape architecture.[9] For many decades, Olmsted was known mainly to historians for his accounts from the 1850s of conditions in the pre–Civil War South, and only since the mid-1970s, through numerous books, articles, conferences, and national attention to his extensive body of work, has he been elevated to the position of "Father of American Landscape Architecture," an appellation sometimes attached instead to Downing. And while Olmsted, due to the length and breadth of his career, may have been the field marshal who established with major projects the profession of landscape architecture in the United States, recent scholarship tells us that there were countless other troops enlisted in related ways and smaller scales elsewhere in America, unseen and unknown

7. Norman T. Newton, *Design on the Land: The Development of Landscape Architecture* (Cambridge: Belknap Press of Harvard University Press, 1971).

8. See, for instance, biographical essays in Charles A. Birnbaum and Robin Carson, eds., *Pioneers of American Landscape Design* (New York: McGraw-Hill, 2000), and Birnbaum and Foell, eds., *Shaping the American Landscape.* For new scholarship on women landscape architects, see Thaïsa Way, *Unbounded Practice: Women and Landscape Architecture in the Early Twentieth Century* (Charlottesville: University of Virginia Press, 2009).

9. See, for instance, David Schuyler, *Apostle of Taste: Andrew Jackson Downing 1815–1852* (Baltimore: Johns Hopkins University Press, 1996), and Judith K. Major, *To Live in the New World: A. J. Downing and American Landscape Gardening* (Cambridge: MIT Press, 1997).

perhaps, but important nonetheless for their regional impact. Certainly Affleck is one such foot soldier.

Curiously, Affleck's diversity of interests—animal husbandry, agriculture, scientific farming, ornamental horticulture, insects, and hydrology, among others—means that several professions could afford him celebrity status, yet none has. If mentioned at all, his contributions are associated with only one area of his accomplishment at the expense of the others. From another perspective, however, it is the diversity of his career interests which suggests that his expertise was rooted in a comprehensive and perhaps unique understanding of the stewardship of the land—how it operated scientifically, chemically, and mechanically; what it contained, organically and inorganically; and how elements such as plants, animals, insects, and humans could be employed for maximum operational efficiency and productivity of the land in a sustainable way. In Affleck's career, as in those of Downing and Olmsted, we find how numerous interests overlap and inspire several modern professions, including landscape architecture, horticulture, animal husbandry, soil science, agricultural engineering, and water management.

In terms of academic investigations, Affleck's career has been examined through a master's thesis and two doctoral dissertations.[10] Another dissertation is in progress at this writing (2014). Outside the existing thesis and dissertations, no detailed biography exists, save the summary essays previously mentioned. A few scholarly articles have appeared over time, and an early one (by Robert W. Williams Jr.) is reprinted here in full.

In the biographical article that follows, Williams notes that, in a letter from 1851, Affleck suggested that agricultural writers such as himself should use the works of others and avoid "wasting their time & labor in endeavoring to discover what others already know."[11] For the biographical narrative of Affleck's life, I have taken that advice, primarily using Williams's article together with Fred Carrington Cole's work to avoid duplicating their efforts but, more important, to bring their scholarship to the attention of today's

---

10. Fred Carrington Cole, "The Early Life of Thomas Affleck" (master's thesis, Louisiana State University, 1936); Fred Carrington Cole, "The Texas Career of Thomas Affleck" (PhD diss., Louisiana State University, 1942); Robert Webb Williams Jr., "The Mississippi Career of Thomas Affleck" (PhD diss., Tulane University, 1954).

11. Affleck to Eli J. Capell, February 6, 1851, cited in Williams, "Thomas Affleck: Missionary to the Planter, the Farmer, and the Gardener," *Agricultural History* 31, no. 3 (July 1957): 42.

audience.[12] Neither Cole nor Williams situates Affleck in the broader context of his times or from the perspective of the profession of landscape architecture, surely a statement of the historiography of *their* times. Nevertheless, Cole's and Williams's works provide the foundation upon which we may consider Affleck's contribution to this region's environmental history through the lens of recent scholarship in the context of his times and the benefit of a broader perspective. In providing annotations to Affleck's writings, I have endeavored to interpret meaning, provide context, and make connections; any misreading or misinterpretations are mine alone.

Tangential to information we gather through representative samples of Affleck's writings are incidental references that add insight to our understanding of the past and its impact on the present. For instance, in Affleck's contribution to the *Southern Medical Reports* (1851) describing the health and management of the enslaved on Mississippi plantations, several examples of his syntax include language still heard today in Mississippi: "comfort" for a quilt or coverlet; "sweet milk" as distinguished from buttermilk or soured milk; "dinner" and "supper" for, respectively, the noon and evening meals. In Affleck's writings we find many people mentioned who are significant community leaders, elected officials, and decision-makers, suggesting that Affleck's circle of acquaintances were those who shaped public opinion and exerted social, political, and economic influence in their communities. Where relevant, I have endeavored to add information enabling the reader to attach a context to the names mentioned. Finally, Affleck's writings, like those of Olmsted and Downing, have a richness and relevance that resonates today, over a century and a half later, because of the simplicity of his style and the depth of information that his writings display.

Affleck's personal correspondence and business archive fill in details of his life and career, and these archives were mined extensively through the works of Williams and Cole. Drawing heavily from that archive, their works favored those sources over his published writings. I propose, however, that it is through Affleck's published writings that we gain a comprehensive appreciation for his place in nineteenth-century American life. Scattered in

12. I am indebted to Professor Clair Strom, current editor of *Agricultural History*, the journal in which Williams's article first appeared, for permission to republish it here, and to Fred Carrington Cole's family for their interest in this project and for permission to extract sections of his works for use here.

newspapers and books that are now obscure and difficult to find, the writings of "Friend Affleck" (as he was often addressed by those with whom he corresponded) have not, until now, been collected.

Some newspapers and books cited as sources for Affleck's writings remain elusive. A comprehensive bibliography of his published writings does not exist and is beyond the scope of this project.[13] Nevertheless, bringing representative examples of Affleck's writings to the foreground will illustrate his relevance in creating a more comprehensive understanding of agricultural and horticultural conditions in the Gulf South region in mid-nineteenth-century America. In the area of agricultural advancements, Affleck's works clearly represent a major force in the dissemination of information, advocacy for scientific methods and new agricultural techniques, and interest in horticultural advancements. These initiatives held great importance for the rural South and provide the context for regional attitudes about land use, environmental resources, and agricultural education.

Emerging nineteenth-century theories that increased awareness of exterior spaces are more apparent in the works and writings of those who lived in urban areas of the East Coast (like Downing and Olmsted), with results more readily apparent in those urban spaces. Yet the focus on agricultural advancements and domestic horticulture that comprise much of Affleck's interests and those of his constituents are obvious evidence of an interest in presenting an attractive, ordered landscape to public view. From that position, it is a short line to an interest in the design of exterior public spaces. Clearly an appreciation for what horticulture can do in domestic and rural situations leads to a willingness to accept similar improvement in the public arena.

In the broader context, I propose that an examination of Affleck's writings and career contributes to a discussion of the origins of the modern practice of landscape architecture. From both, we can see how the profession of landscape architecture has roots throughout the country beyond East Coast communities and personalities such as Downing and Olmsted.

13. Cole mentions (in "Texas Career," 18 n. 23) that a "complete bibliography of Affleck's published writings is being compiled by the author and Professor Stephenson." Thus far, this document, if it exists, has not been found. The reference is to southern historian Wendell Holmes Stephenson (1899–1970), professor of history and dean at LSU (1927–1942) and editor of the *Journal of Southern History* from its founding at LSU in 1935 until 1941. Stephenson was Cole's academic advisor and mentor. See Thomas D. Clark, "Wendell Holmes Stephenson, 1899–1970: Master Editor and Teacher," *Journal of Southern History*, 36, no. 3 (August 1970): 335–49.

Affleck's career took him to the American South and to writing for publications now forgotten and sometimes also lost. What remains of his writings, however, demonstrates his encyclopedic interests and the broad range of his knowledge. Through his writings, we can situate him in the broader picture of American environmental design as a regional pioneer whose works played a part, even if small, in the development of landscape architecture in America. My hope is that these writings—as Affleck's obituary notes, the "record he has made of his greatness"—will enable us after over one hundred and fifty years to appreciate his important contribution to the environmental heritage of the Southwest of mid-nineteenth-century America.

# EDITORIAL NOTE

Given here are the works of four hands: first are biographical essays, the academic works of two scholars from 1942 and 1957; and second are representative examples of Affleck's writings from various nineteenth-century agricultural newspapers and books. Finally, my introductions and annotations are intended to provide the twenty-first-century reader with the historical context in which to place Affleck's writings and those who wrote about him.

Affleck's articles, for the most part, appear in their original forms, but I have summarized or abbreviated passages when I felt they have limited relevance for today's audience. Selections from the *Western Farmer* are taken from the second printing of the 1850 hardbound edition of collected articles that originally appeared in newspaper format, and I have assumed that what appears in this collected edition represents what the editors considered significant from the newspaper's recent past. Throughout, I have silently corrected idiosyncratic punctuation and syntax, made minor spelling adjustments and typographical corrections when obvious infelicities occur or when such corrections clarify meaning, and kept Affleck's sometimes odd use of italics for emphasis within his text. The academic works of Williams and Cole are given as they first appeared; citation formats in their works do not coincide with contemporary ones employed here. Because I have inserted a few clearly identified notes of my own into their material, the numbering of their citations has changed, but otherwise their notes are reproduced exactly where and as they appeared in the originals.

My contributions appear as introductory or concluding comments and editorial observations. Annotations appear as footnotes or within brackets. Overall, they are intended to provide supplementary information, background, and a contemporary context to facilitate the reader's insight through access to current research and the hindsight of history.

Scattered throughout, the reader will find woodcut images representative of those commonly seen in nineteenth-century agricultural publications, and some are associated with Affleck himself. For instance, the escutcheon with the staff of wheat and the phrase *Preciosum quod utile* (What is useful is precious) on the first page of this book appears initially on the title page of the 1851 edition of *Affleck's Southern Rural Almanac;* it also appears on subsequent editions in the 1850s. Affleck registered the almanac's title, together with this symbol, with the United States District Court, Southern District of Mississippi, on August 25, 1851, for the 1852 edition, "in conformity with the act of Congress . . . respecting Copyrights," and a copy of his application is found in the Affleck Collection at LSU's Hill Memorial Library. What appears on the title page for Part II, the Selected Writings, is from an advertisement for F. D. Gay's Seed Store and Plant Depot of New Orleans, found in *Affleck's Southern Rural Almanac* of 1851; Affleck and Gay enjoyed a lengthy and fruitful personal and professional relationship, and advertisements for Gay's business appear often in Affleck's *Almanacs.* Finally, the image of the grapevine on page 47 accompanied Affleck's article "The Grape" in the *Western Farmer and Gardener.* Other images, some generic, others specific to articles they illustrate, have come from nineteenth-century sources such as *The Cultivator* (Vol. I, 1844).

Acknowledgment is due to graduate assistants Claudio Golombek and Kossen Miller, who assembled sources and checked facts. Finally, I owe a note of thanks to Debbie de la Houssaye, without whose support this book would not have been possible.

# I

## A BRIEF BIOGRAPHY

—— OF ——

# THOMAS AFFLECK

The major biographical narrative of Thomas Affleck's life comes from works of Robert Webb Williams Jr. (1923–93) and Fred Carrington Cole (1912–86), and it is from their academic scholarship that the following selections come.[1] Their academic investigations of Affleck, completed between 1936 and 1954, supply detailed information about three phases of Affleck's life and provide us with perspectives through the lenses of the authors' times and assessments of his life and career, six to eight decades after his death. Both scholars approach Affleck as a person of historical importance in the nineteenth-century South. Neither, however, discusses a national context of Affleck's career within the subject areas of his writings; neither identifies connections between Affleck and others writing about and advocating for similar subjects; and neither speculates on the importance of Affleck in advancing contemporary interest in ornamental horticulture or the design of exterior spaces in the geographical areas in which he lived and worked. Their approaches are more indicative of the historiography and the resources of their times rather than of a deficiency of scholarship, and it should be noted that scholarly appraisals of the lives and careers of Andrew Jackson Downing and Frederick Law Olmsted, the national figures with whom I argue Affleck should be considered, did not appear until much later in the twentieth century.[2]

1. Williams, "The Mississippi Career of Thomas Affleck"; Williams, "Thomas Affleck: Missionary to the Planter"; Cole, "The Early Life of Thomas Affleck"; and Cole, "The Texas Career of Thomas Affleck." The article given here by Williams, drawn from his dissertation, appeared in 1957.

2. See, for instance, Schuyler, *Apostle of Taste,* and Major, *To Live in the New World.* For Olmsted, see Laura Wood Roper, *FLO: A Biography of Frederick Law Olmsted* (Baltimore: Johns Hopkins University Press, 1973).

The biographical works of Robert Webb Williams Jr. and Fred Carrington Cole, which form the foundation for a twenty-first-century appraisal of Affleck, when combined with Affleck's own words, facilitate a comprehensive assessment of an important agricultural reformer, a passionate advocate for scientific farming, agricultural education, and horticultural advancements, and a visionary interested in numerous subjects ranging from beekeeping, animal husbandry, and antebellum agricultural conditions in Mississippi to labor and immigrant issues, shipping, and the economic development of post–Civil War Texas.

The first essay is an article by Williams that appeared in 1957 in *Agricultural History* 31, no. 3 (July 1957), the journal of the Agricultural History Society, and appears here with permission from its present editor.[3] Obviously this article evolved from Williams's unpublished dissertation at Tulane University, "The Mississippi Career of Thomas Affleck" (1954). Both Williams and Cole acknowledge that this period of Affleck's career was his most influential, and certainly these years were his most productive in terms of publications. The second essay, composed of material taken from Cole's 1942 LSU dissertation, summarizes earlier events in Affleck's life and discusses his Texas years, from 1857 to his death in late 1868.[4] The works of both men refer to printed works but are notable for the rich detail resulting from their exhaustive readings of Affleck's correspondence.

Both Williams and Cole were academics, and combined, their academic works exceed eight hundred typewritten pages with hundreds of citations. Williams later published articles on historical figures in Louisiana and Texas and edited the letters of Affleck's son Isaac, written during his service in the

3. "The Agricultural History Society was founded in Washington, DC in 1919 'to promote the interest, study and research in the history of agriculture.' Incorporated in 1924, the Society began publishing a journal, *Agricultural History*, in 1927. The term 'agricultural history' has always been interpreted broadly, and the Society encourages research and publishes articles from all countries and in all periods of history. Initially affiliated with the American Historical Association, the Agricultural History Society is the third oldest, discipline-based professional organization in the United States. Currently the membership includes agricultural economists, anthropologists, economists, environmentalists, historians, historical geographers, rural sociologists, and a variety of independent scholars." See http://www.aghistorysociety.org/society/, accessed July 24, 2012.

4. In his work, Cole acknowledges the direction of his advisor, Dean Wendell H. Stephenson, and James A. McMillen, through whose efforts the Affleck Papers came to the LSU Archives.

Civil War.[5] At the time Williams's article was published, he was a professor of social sciences at East Carolina College in Greenville, North Carolina. Cole, after taking degrees from LSU, moved into administration, first at Tulane University then as president of Washington and Lee University in Lexington, Virginia, from which he retired in 1967. Cole did not expand on Affleck beyond his thesis and dissertation, although his unpublished manuscript, "Thomas Affleck and Ante-Bellum Agriculture" (December 28, 1940), is in the Affleck Papers, Cushing Memorial Library, Texas A&M University, with this notation: "Paper presented at the American Farm Economic Association meeting in New Orleans as a Joint Program with the Agricultural History Society, December 28, 1940." Family members do not recall mention of Affleck, and selections from Cole's dissertation appear here with the gracious permission of his son.[6]

These biographical works, drawing heavily on Affleck's personal and business papers, are the foreground for an examination of Affleck's published works. Together, they situate him as a regional participant in mid-nineteenth-century American movements advocating for agricultural reform, increasing horticultural awareness, and encouraging aesthetic appreciation among the general public. Examining these environmental attitudes through Affleck and his writings adds a regional perspective to a larger understanding of the origins of the profession of landscape architecture in America.

5. See Robert W. Williams Jr. and Ralph A. Wooster, eds., "With Wharton's Cavalry in Arkansas: The Civil War Letters of Private Isaac Dunbar Affleck," *Arkansas Historical Quarterly* 21, no. 3 (Autumn, 1962): 247–68; and Robert W. Williams Jr. and Ralph A. Wooster, eds., "Camp Life in Civil War Louisiana: The Letters of Private Isaac Dunbar Affleck," *Louisiana History* 5, no. 2 (Spring 1964): 187–201.

6. See Cole's biography as past president of Washington and Lee University: http://www.wlu.edu/presidents-office/about-the-presidents-office/history-and-governance/past-presidents/fred-carrington-cole, accessed September 9, 2013. Communications with Cole's son, R. Grey Cole, October 7 and 11, 2011).

# 1

## Thomas Affleck: Missionary to the Planter, the Farmer, and the Gardener

ROBERT W. WILLIAMS JR.

*Agricultural History* 31, no. 3 (July 1957)

---

*[Editor's Note: Most of the footnotes below are Williams's, original to the article. I have added a few comments for clarity or additional context, which are indicated by "—Ed."]*

---

Twenty years before the creation of the United States Department of Agriculture and three decades before the establishment of the first agricultural experiment station, Thomas Affleck of Washington, Mississippi, assembled, collected, and presented to the farming public what was [*sic*] among the best and most advanced practices of his day. The letters and published works of this much-neglected figure of American agricultural history cover a variety of subjects, from the layout of a kitchen garden to the management of an entire plantation; from the care of house cats to the need for the development of southern industry. Many of his projects and enthusiasms extended over a period of several years. He became a leading expert in a number of fields and one of the outstanding publicists of agricultural improvement in the Southwest. The content of his articles, letters, and publications and the fact that his opinions were well received marked him as one of the movement's most effective advocates.[1]

Affleck was born in Dumfries, Scotland, in 1812, the son of a general merchant and importer. He came to America in 1832, spent eight years in New York, Pennsylvania, Indiana, and Ohio employed as clerk, merchant, and small-scale nurseryman before he accepted a position as associate editor of

---

1. James T. Adams and R. V. Coleman, eds., *Dictionary of American History* (6 vols., New York, 1848), 1:25, 26.

the *Western Farmer and Gardener* of Cincinnati in 1840.[2] From the beginning of his association with the *Western Farmer and Gardener,* and particularly after he became sole editor in 1841, he widened the scope of the magazine to include such departments as livestock, bee breeding, horticulture, and viticulture, as well as the usual columns devoted to staple farming. He advocated agricultural societies, agricultural colleges, and horticultural groups. He pointed out the evils of a one-crop economy to both the West and the South. He expressed his ideas of what constituted an ideal farm—a place to provide year-to-year security, a place of beauty and satisfaction, as well as a source of profit.[3]

Livestock became known as Affleck's chief interest and competence, and he took advantage of his reputation to extol the virtues of several animal breeds. Partly in the hope of making a profit from the artificially created demand for these animals, he gathered together a herd of selected stock in 1841 for a trip to the South. With an eye directed toward both reader-appeal and profit, he intended to tap the vast store of interest being generated in the lower South. Agricultural societies encouraged improved breeding, while the most scientifically advanced planters and farmers banded together to import prize specimens of cattle, sheep and hogs.[4] With these favorable signs and despite the known preference of many Southern farmers for their own scrub animals,[5] he set forth to "stir up and nourish a taste for the improvement of the common stock." He intended also to report on the "much belied" South-

2. Information on Affleck's early life is found in an account kept by one of the present Affleck family on a trip to Scotland in 1935. This account and all other manuscript material cited herein, unless otherwise indicated, is in the Affleck Papers, Louisiana State University, Department of Manuscripts and Archives, Baton Rouge, Louisiana. The early life of Affleck is treated in detail in Fred C. Cole, "The Early Life of Thomas Affleck," master's thesis, Louisiana State University, 1936. See also: Thomas Affleck to Andrew Hannay, July 23, 1832; Affleck to John Graham, August 16, 1832; Diary of Affleck, 1832, *passim.*

3. *Western Farmer and Gardener,* 2: passim (1840–41).

4. Ibid., 3:78–85 (1842); Charles T. Leavitt, "Attempts to Improve Cattle Breeds in the United States, 1790–1860," *Agricultural History,* 7:51–67 (April, 1933); Thomas D. Clark, "Livestock Trade Between Kentucky and the South, 1840–1860," *Register of the Kentucky State Historical Society,* 27:569–581 (September, 1929); *Agriculture in the United States in 1860: Compiled from the Original Returns of the Eighth Census, Under the Direction of the Secretary of the Interior* (Washington, 1864), 112.

5. *Western Farmer and Gardener,* 3:78–85 (1842); *Southern Cultivator,* 1:127 (1843), 2:183 (1844), 3:121 (1845); *Southern Agriculturist,* n.s. 2:521 (1842); *De Bow's Review,* 3:2 (1847). This was only one phase of the considerable prejudice against experimental farming and stock-breed-

ern plantation economy and to make new friends for his journal.[6] At Vicksburg, his first stop, he sold a part of his stock and then went on to Natchez and Washington, Mississippi, to attend the annual fair of the Agricultural, Horticultural, and Botanical Society of Jefferson College.[7] After the fair he traveled and re-traveled the water and land routes between Natchez, Baton Rouge, Vicksburg, and Jackson, and made a short visit to New Orleans. Not until February of 1842 did he return to Cincinnati, and then only to terminate his affairs so that he might make a permanent removal to the South.[8]

During his stay in the Natchez area, he met and married a young widow he supposed to be a wealthy landholder, but after his removal to Mississippi he found that her properties, though extensive, were heavily encumbered by debt. Bad growing seasons, low prices, and debts reduced further the holdings he had hoped to operate and forced him to turn from large-scale cotton planting to commercial nursery operations, writing, publishing, and promoting. None of these endeavors was financially profitable, but in each field he rendered a worth-while service to the farming community of the Southwest.

The financial accounts of Affleck's plantation operations indicted him as a failure, and the court records of Adams County strengthened the charge, as they traced the gradual loss of the property entrusted to his management. But if he was an unsuccessful cotton planter, so were many others in the difficult years of the 1840s and most of them were marked for oblivion. Affleck, however, whose arrival in Mississippi was almost coincidental with

---

ing during this period. It was often cited that agricultural leaders were poor farmers. *Farmers Register,* 2:17 (1834); Lewis C. Gray, *History of Agriculture in the Southern United States to 1860* (2 vols., Washington, 1933), 789.

6. *Western Farmer and Gardener,* 3:78–85 (1842). It was a common practice for editors to travel and to report their trips for the pages of their journals. Sometimes they served as agents for their magazines and for seed houses, or implement manufacturers. Southern editors and reporters travelling in the North often carried commissions to purchase machinery and stock for their readers and neighbors. Albert L. Demaree, *American Agricultural Press, 1819–1861* (New York, 1941), 106–107.

7. *Western Farmer and Gardener,* 3:78–85 (1842). See also Charles S. Sydnor, *A Gentleman of the Old Natchez Region, Benjamin L. C. Wailes* (Durham, 1938), 80, 152. There was a great deal of local pride in Jefferson College which had been incorporated by the legislature of the Mississippi Territory on May 13, 1802. J. K. Morrison, "Early History of Jefferson College," *Publications of the Mississippi Historical Society,* 2:179–188 (1899).

8. See letters exchanged between Affleck and Anna M. Smith in the months of February and March, 1842.

the spread to the Southwest of the movement for agricultural improvement, shares the credit for the dissemination of advanced agricultural methods with men like Noah B. Cloud, Martin W. Philips, and Solon Robinson.

As publisher, editor and voluminous letter-writer, he became a missionary to the planter and the farmer. In his published articles and in private correspondence he answered the questions of his fellow farmers. The increasing number of correspondents who sought his advice and the thousands who read his comments and suggestions did not condemn him as a failure on the strength of evidence presented in terms of mere dollars and cents.

Affleck became agricultural editor of the New Orleans *Weekly Picayune*[9] in January of 1851.[10] The columns that he wrote for the *Picayune* were well received throughout the lower South. The advice he gave in answer to queries was usually good, resulting as it did from wide reading, experimentation, and a clear, concise language which reminded one reader "more . . . of Cato that any that I ever read."[11] Affleck did not strive for originality. He adopted the view that agricultural writers should avoid "wasting their time & labor in endeavoring to discover what others already know."[12]

He considered it a service to pass on information of use to southern farmers gathered from sources that they would not ordinarily see. He thought it the duty of every farmer to subscribe to at least two agricultural publications and pay their fees promptly, but he did not expect them to be able to read and evaluate all of the available material.[13] This he accepted as his responsibility.[14] To this end, he owned what was probably one of the largest agricultural libraries in the Southwest, including full files of all the major agri-

9. [*There was a* Weekly Picayune, *of which few copies exist, and a daily* Picayune, *easily found, beginning in 1837. In the 1840s it became the* Daily Picayune. *After a series of mergers and consolidations, it later became the* Times-Picayune.—*Ed.*]

10. Affleck to A. M. Holbrook, January 6, 1851.

11. Robert Russell, *North America: Its Agriculture and Climate* (Edinburgh, 1857), 254. Russell, in touring the South, had made inquiries for Affleck but met him by chance in the lobby of the St. Charles Hotel in New Orleans. Russell was pleased to find that Affleck was a Scotsman. He wrote that "of all the parties to whom I had introduction in the United States, no one, somehow or other, did I consider I had so great a claim to attention as upon the one whom I had so unexpectedly met."

12. Affleck to Eli J. Capell, February 6, 1851, in Eli J. Capell Papers, Louisiana State University, Department of Archives and Manuscripts, Baton Rouge, Louisiana.

13. *Prairie Farmer*, 4:106–107 (1846); Affleck to B. M. Norman, January 6, 1851.

14. Affleck to B. M. Norman, January 18, 1851.

cultural and horticultural journals printed in the United States and a great many of the leading English and European works.[15] He frequently ordered books from New York, Cincinnati, England, and the Continent.[16] He tried to obtain the latest pronouncements, as well as the standard works, on agriculture, horticulture, and the natural sciences.[17] Upon accepting his appointment at the *Picayune*, he requested that the editor arrange exchange subscriptions so that he might have ten American and six English magazines in addition to those he regularly received. He thought that this was essential to enable him to "keep up with the times [and] to do the department justice."[18]

Not only did he rely on books and periodicals, but he made constant use of the material that came to him in letters. Correspondence with experts in several different fields gave his column an encyclopedic character. He published excerpts from letters of specialists and ordinary farmers who wrote to him telling of their experiences and their discoveries.[19] He often sent out letters specifically to elicit information for his column, sometimes circularizing a part of the farm population on crops, methods, implements, and other rural matters. He tried, without success, to get the editor of the *Picayune* to bear a part of the expense of sending out 33 questionnaires on Southern Agriculture. He argues that in addition to furnishing him information it would increase circulation, since those who stated opinions would be interested to see the consensus report.[20]

He encouraged his readers and his correspondents to experiment and to report their findings to him for evaluation, collation with his own knowledge

---

15. List of books and magazines owned by Affleck in Memorandum Books 1852–53; Affleck to A. Hart, published, May 10, 1851; Affleck to Mr. Warner, January 19, 1854; Affleck to D. Redmond, July 16, 1854; Affleck to Gales and Seaton, March 1, 1856.

16. Affleck to B. M. Norman, January 18, 1851. Affleck had great respect for the opinions of the Scotch [*sic*] on gardening and the English on the subject of evergreens, grasses, and ornamental shrubs. Affleck to D. Redmond, February 9, 1854; January 14, 1854; Affleck to J. C. Morgan, May 28, 1851.

17. Affleck to A. M. Holbrook, January 16, 1851; Affleck to F. D. Gay, May 13, 1851; Affleck to Morton and Griswold, March 10, 1854.

18. Affleck to A. M. Holbrook, January 16, 1851.

19. See, for example, his writings in the New Orleans *Weekly Picayune* during 1850–51. He also wrote articles on a fairly regular basis for the New Orleans *Commercial Times* during 1846, and on an occasional basis after that period.

20. Affleck to A. M. Holbrook, May 13, 1851. Affleck had sent out circular letters before but decided that he could no longer bear the expense alone.

and experimentation, and finally for publication, so that useful conclusions might be put to wider use.[21] In his column, he often noted new books that he received from the publisher, and was particularly pleased to recommend those which were pertinent to reforms that were needed in southern rural life.[22] By a variety of methods, Affleck was able to assemble and present advanced advice and counsel. In the 1850's, he advocated many practices and procedures of good farming through his weekly column that would receive the approval of present-day county agents and state experiment station personnel.

Affleck offered diversification as a sound practice for the plantation and the South and was willing to frame his recommendations in detail. He encouraged experimentation with crops that might supplement cotton. He admitted that many new ventures might not pay any better than cotton at first and that experimentation was costly. He advocated government subsidies for crop research, but in the absence of such support, he performed a great service by publishing the results of his own experiments and by giving publicity to the findings of others. He tried to convince his neighbors that much of the energy that they spent cultivating cotton could be better employed in raising livestock, developing new crops, finding new uses for old crops, and providing more of the foodstuffs required on the plantation.[23]

Affleck felt a moral obligation to posterity, as well as an economic compulsion to preserve the land, and did much of his most effective work in the field of conservation. To prevent erosion, he advocated contour plowing. To restore plant food to the soil, he suggested measures which were in accord with the most advanced principles of agricultural chemistry, including systematized rotation and the liberal use of both natural and commercial fertilizers. He was similarly interested in saving each year's crop from its natural

21. New Orleans *Weekly Picayune,* April 24, 1850. During a few months of 1850 and sporadically before that time Affleck had contributed articles on a volunteer basis. He hinted to the editor that some remuneration might be in order for this volunteer work before he was put on the regular payroll but there is no indication that he ever received any money for these services.

22. Affleck to A. Hart, May 10, 1851. Affleck, in acknowledging Kingsford's work on plant roads, told the publisher that he would give it a prominent and favorable notice even though there was no new material in the book that had not already been published in agricultural journals because it would "serve my purpose well in stirring up my neighbors on the subject."

23. *American Agriculturist,* 3:28–31 (1845); *Norman's Southern Agricultural Almanac* (New Orleans, 1847), 21.

enemies, disease and insects. He approached the problem in the same manner that is now deemed essential by professional entomologists.[24]

Affleck was outstanding as a nurseryman and counselor to the gardener. There were no large nurseries available in the lower South in the 1840's and the excessive cost and uncertainties of transportation discouraged purchases from the north. The plants that Southerners did order often proved to be unsuitable for Southern growing conditions. Most of the Southern population was suspicious of Northern nurserymen because of repeated failures with Northern stock. Affleck offered both information and stock adapted for Southern use. His services were generally accepted without prejudice since he was known to be a Southerner devoted to the interest of his section. He introduced many kinds of evergreens, shade trees, flowers, and hedges that added beauty to the Southern scene and several types of fruit trees that added variety to the Southern diet. He made costly importations from England and the Continent and spent years acclimating plants to the Southern soil and climate. Since most Southerners lacked any appreciation or knowledge of horticulture and since there were few treatises or journals dealing with Southern conditions, Affleck found it necessary, not only to ready stock for sale, but also to acquaint the public with the names and uses of the varieties and the proper methods of selection and cultivation.[25]

By 1850, Affleck had gained national reputation and stature as an expert consultant on agricultural and horticultural matters. He was sought as a speaker and for committee work by organizations devoted to advancement in both fields. He earned the gratitude and respect of farmers in all parts of the Union; his advice was constantly sought on a great variety of agricul-

24. Affleck to President of the General Society of Agriculture of France, undated; undated manuscript in the Affleck Collection entitled "On the Southern Cow-pea"; *American Agriculturist,* 2:17 (1843); 3:181–82 (1844); *Affleck's Rural Almanac and Plantation and Garden Calendar, 1851,* 6, 41–44; *1853,* 3. This will hereinafter be cited as *Almanac.* Affleck to B. M. Norman, May 1, 1851; Edmund Ruffin to Affleck, July 10, 1844; John Pitkin Norton to Affleck, September 29, 1845; Affleck to Noah B. Cloud, March 4, 1853; *Southern Agriculturist,* 8:47 (1834); Affleck to Editor, *New Orleans Commercial Times,* October 2, 1847. See also Affleck's extensive correspondence with Thaddeus W. Harris of Harvard University, Lewis W. Harper, entomologist of Savannah, Georgia, and Professor Nicholas Hentz of the University of North Carolina during the 1840's.

25. *Almanac, 1852,* 91–108, *1855,* 47; Affleck to E. Zwilchenbart and Company, March 13, 1851; Affleck to William Skirving, March 13, 1851; Affleck to Edward Bossange, November 27, 1842; Natchez *Daily Courier,* October 28, 1854; Affleck to Thomas Rivers, September 14, 1852; *Southern Cultivator,* 10:42 (1853)

tural and sectional problems. Almost every mail brought a flood of questions which Affleck considered carefully and answered courteously, adding to his long list of friends, widening his influence and his opportunity to serve further those who made their living from the soil.

Thomas Affleck claimed to be the originator of a new type of almanac different from the "comic absurdities" that flooded the market before he issued the 1842 *Western Farmer and Gardener's Almanac.*[26] In 1846, he began to edit an annual series of almanacs published by B. M. Norman called *Norman's Southern Agricultural Almanac.*[27] In 1851, the name was changed to *Affleck's Southern Rural Almanac and Plantation and Garden Calendar.*[28] A primary purpose of the later publications was to advertise Affleck's Southern Nurseries but its usefulness transcended its service as a catalogue of trees and plants. The farmers of the Southwest adopted it as a guide to both new and accepted practice.[29]

The Almanac presented charts and calendars to serve as guides for planting, cultivating, and gathering crops and furnished tables of standard weights and measure. There were rules for land conservation, restoring plant food to the soil, and combatting insect pests, plant diseases, and weeds. Long and authoritative articles on the proper care of orchards, flowers, vegetables, and ornamental shrubs constituted a most valued section of the Almanac and were intended to mesh with Affleck's efforts to stimulate sales

26. *Norman's Southern Agricultural Almanac,* unnumbered first page of introduction. Although Affleck was proud of the 1842 edition of the *Western Farmer and Gardener's Almanac,* he expressly denied any responsibility for the 1843 edition which listed his old partner, Charles Foster, as editor. See first page of the preface of the 1843 edition of this work.

27. B. M. Norman was a prominent and influential publisher and book dealer in New Orleans.

28. A copy of the 1851 copyright is in the Affleck Papers.

29. The few short-lived magazines of the section had a small circulation and little lasting influence. The *Southwestern Farmer,* of Raymond, Mississippi (1842–1845), and the *Southern Planter* of Natchez, Mississippi, which lasted for only one year (1842), were typical of the Southwestern agricultural press. Demaree in his *American Agricultural Press* lists no important agricultural magazine generally circulated in the Southwest and Frank L. Mott makes no mention of any such magazine in his *History of American Magazines* (New York, 1930). Planters like Eli J. Capell and M. W. Philips, and publishers like J. D. B. De Bow and B. M. Norman recognized the need for a strong agricultural journal in the Southwest and tried without success to establish one. Eli Capell to Affleck, November 15, 1846; *De Bow's Review,* 5:82–86 (1848); Affleck to B. M. Norman, January 18, 1851.

for his own nursery. Seldom did he overlook a chance to air his view that trees and plants, to be grown successfully, should be carefully acclimated to the South. There was usually the suggestion that this valuable process was best accomplished at his own Southern Nurseries. The work that should be accomplished in each month was set down under the heading of cotton plantation, sugar plantation, kitchen garden, plantation garden, flower garden and shrubbery, fruit garden, and orchard.[30]

The publication of Affleck's *Plantation Record and Account Book* for cotton and sugar growers was a result of Affleck's experience as a cotton planter. Upon arrival in Mississippi, he found that it was necessary because of the locations of the plantations he controlled, to rely on overseers. The overseers in his employ were "ignorant of everything like accounts & in fact, system of any kind." Since they "could not be relied upon to draw up their report after their own fashion," Affleck indicated by blank forms the type of information that he required. B. M. Norman, while visiting in Adams County, saw the forms and suggested that they might be published. Affleck adopted Norman's suggestion and prepared for publication record books designed for cotton and sugar plantations of varying size. The first editions published by Norman in 1847 sold well without too much effort on the part of the publisher.[31] Affleck's nursery affairs and various other considerations prevented a revised edition the year following, but in subsequent years until 1860, Af-

[*Williams is correct that agricultural magazines and newspapers published in the South were generally short-lived, but suggesting that they had "little lasting influence" can be argued. Several journals should be mentioned as evidence of efforts across the South, beginning as early as 1828 and increasing in the 1840s and 1850s, when Affleck was actively promoting his* Almanac, *to generate interest in such agricultural publications, including the* Southern Agriculturist *(Charleston, South Carolina, 1828-46); the* Tennessee Farmer *(Jonesborough, Tennessee, 1836–40); the* Southern Cultivator *(Georgia, 1843–1935); the* American Cotton Planter *(1853–61); and the* Mississippi Planter & Mechanic *(Jackson, Mississippi, Vol. I, 1857). Demaree's encyclopedic study mentions all but the* Mississippi Planter & Mechanic. *It is likely that Williams did not have access to copies of these journals for first-hand observation and relied instead on information given by Mott and Demeree. The lack of access by scholars of the past to these journals is now alleviated through online resources.—Ed.*]

30. The above brief outline is merely a summary of the material in all of the *Almanacs* from 1847 to 1860.

31. Affleck to Secretary of Executive Committee of the Southern Central Agricultural Society of Georgia, May 24, 1851. The types of books published for cotton plantations were: Number 1 for

fleck published and sold copies of these books introducing them into every state of the South.[32] Yearly revisions incorporated some suggestions from overseers and planters, but most of the changes were not radical.[33]

The *Plantation Record Books* brought together in one place several of the miscellaneous records usually kept by Southern staple producers. The record of daily events, slave lists, and stock and implement inventories were included, but so also were a number of other forms which marked an improvement in the system of rural bookkeeping.[34] The record forms were essentially consistent with the intent and purpose of modern cost-accounting, and followed the best and most advanced principles of efficient administrative management.[35] The record books considered the often neglected factors of capital depreciation, labor costs, and social welfare.

Perhaps because of publication troubles he experienced with the *Almanac* and the *Plantation Record and Account Book,* Affleck was never able to bring to completion some other projected works. Five years before he began to write a column for the *Picayune,* Eli J. Capell, a greatly respected Mississippi planter, suggested that he join with Dr. Martin W. Philips to edit a journal for farmers in Mississippi. Capell believed that it would be well-received since there were enough farmers in the state to support it and enough advanced agriculturists to provide material for its pages.[36] Affleck, who had

---

40 hands or less at $2.50; Number 2 for 80 hands or less at $3.00; and Number 3 for 120 hands or less at $3.50. In 1860 he published a Number 4 book for over 120 hands. *Almanac, 1860,* 3. For sugar planters there was a Number 1 book for 80 hands at $3.00 and a Number 2 book for 120 hands at $3.50. Affleck to George Heroman, April 21, 1851. The forms were different for the sugar books. There were spaces for such entries as "strength of juice" and "quantity of lime used in each strike." Provision was made to record the number of hogsheads made each day and the number of cords of wood used, and the inventories of sugar house machinery, tools, and equipment. Affleck intended to compile an account book for non-staple farming but never found time to complete such a work. Affleck to B. M Norman, June 6, 1851.

32. Affleck to B. M. Norman, October 16, 1852; *American Agriculturist,* 6:346.

33. *Almanac, 1853,* 72–73; Affleck to Governor J. H. Hammond, January 3, 1855. [*Hammond was governor of South Carolina from 1842 to 1844.—Ed.*]

34. Gray, *History of Agriculture in the Southern United States to 1860,* 543; *Farmer's Register,* 4:725–27 (1836).

35. John L. Stone, "A Simple System of Common Farm Accounting," *Cyclopedia of American Agriculture,* 4:261–232; J. S. Hall, "Farm Inventories," U.S. Department of Agriculture, *Farmer's Bulletin* (1920), 1182.

36. Eli J. Capell to Affleck, November 18, 1851.

always expressed a reluctance to invest money in such a venture, did not consider the prospect seriously.[37] His success as an agricultural columnist brought other proposals. James D. B. De Bow suggested that Affleck join him in the publication of a weekly agricultural magazine.[38] In 1848, Affleck had suggested to De Bow the need for a monthly journal of agriculture published in the Southwest to "help dispel ignorance and prejudice against theorizing and book-farming."[39] In 1850, he had decided to publish a magazine to be called the *Southern Rural Magazine* which was to be issued bi-monthly in numbers of 80 pages. Affleck proposed an agreement with a publishing company in which they were to assume all "duties and incur risks and liabilities of publishers." Affleck was to perform all the duties of editor. The net profits were to be divided at the close of each volume.[40] It was still his opinion in 1851 that "we must have a monthly journal of agriculture, either connected with the review, forming a part of it, or as a separate work." Affleck was opposed to a weekly journal, however, and still was unwilling to assume any part in such an enterprise except as an editor on salary. "If I edite [*sic*] a work of this kind," he declared, "I must be reasonably well paid and paid surely and regularly."[41]

In the same year that he began his services as agricultural editor of the *Picayune,* he was engaged with a "somewhat elaborate work on cotton-planting." In this work he would treat the "entire cotton plantation from the cutting of the first stump to the shipping of the Bale."[42] He hoped also by the winter of that year to publish a "large book on southern gardening—Fruit flowers and Vegetables—which will be very saleable if well canvassed."[43] In 1852, he felt compelled "to defer until next summer," his work on the garden.[44] By fall of 1856, he was "copying off for the press, my little volume on the Garden" and hoped to have it ready "in three or four weeks." He estimated that it would be about 300 pages in length and would sell for $1.[45] There are pres-

37. Affleck to B. M. Norman, January 18, 1851.
38. Affleck to James D. B. De Bow, January 23, 1851.
39. *De Bow's Review,* 5:82–86 (1848).
40. Proposed Agreement with Weld and Company by Thomas Affleck, June 5, 1850.
41. Affleck to James D. B. De Bow, June 23, 1851; Affleck to A. Hart, February 17, 1851.
42. Ibid.
43. Affleck to J. Louis Jourdan and Company, August 14, 1851.
44. Affleck to C. G. Edwards, November 14, 1851.
45. Affleck to J. C. Morgan, September 25, 1856.

ent in the Affleck Collection some undated manuscripts which are very likely chapter drafts of Affleck's two projected works, but none of this material was ever published in book form.[46]

Affleck sought ways to combine his own interest with that of the farming community and, at times when he needed cash badly, he was less likely to consider the needs of his fellow farmers than he was the possibility for ready profit. He was keenly conscious of the advantage of his position as a recognized agricultural consultant. He appreciated the power of publicity, the attraction of novelty, and the infallibility often attached to the printed word. Affleck accepted the salary of "something over $1000 per annum" as agricultural editor of the *Picayune,* even though it was less than he had asked, "as it serves my purpose well, to do so."[47] He knew that the position would increase his stature as an agricultural expert. He intended to call attention to his nursery and his publications through the *Picayune* columns, to promote occasionally some item in which he was financially interested, to give favorable notices to various products in return for direct compensation, or as a friendly gesture toward a friend or business associate. "As agric. Ed. of the Picayune," he told a pecan grower seeking an agent in the Southwest, "I have much in my power in the way of aiding a cause of this kind."[48]

In his search for profit, he followed a number of courses. He publicized and sold what belonged to him, he promoted the interests of others for a direct compensation or on a commission basis, and he acted as an agent for a variety of products, ranging from sewing machines to life insurance. Through his regular columns in the New Orleans *Weekly Picayune,* his articles contributed to the New Orleans *Commercial Times,* his articles contributed elsewhere, and through the pages of the *Southern Rural Almanac,* he often recommended plants, livestock, and products in which he had a pecuniary interest. He praised slate roofs in the hope of receiving such a roof for his own house at a reduced price. He was willing to make favorable notice of a printing press, provided the manufacturer purchase advertising space in his *Almanac.* He would use whatever language the manufacturer wanted if the proposed notice were drafted so as not to resemble too closely an advertisement. "Everything of the kind that I do, must be done cautiously

46. A large body of this material in the form of undated manuscripts is located in the Affleck Collection.

47. Affleck to B. M. Norman, January 18, 1851.

48. Affleck to Dr. Junius Smith, June 22, 1851.

& prudently," he warned, "else the columns of the Pic: might be closed to any notices at all."[49]

Affleck was willing to create a market for a product but he was also glad to benefit from a demand already existing. By the 1850's, most farmers were convinced of the beneficial effects of some type of fertilizer, and many were willing to experiment with several. In 1851, Affleck made an arrangement to buy the refuse left from the manufacture of soda water by a New Orleans firm for $1.50 per hogshead. He used some of it as fertilizer at his nursery and sold the rest through a business associate in New Orleans, F. D. Gay. The residue was a saturated sulphate of lime and preferable to the natural article as a provider of plant food.[50] Affleck called attention to this fact in the various publicity outlets available to him, and managed to receive some return for his trouble.[51]

The soda water manufacturer realized that he had been selling a valuable product for less than its true worth and demanded a larger share of the returns. Affleck felt little confidence in the general public's ability to think for itself and threatened to reverse his field and decrease the demand as he had accelerated it. He instructed his New Orleans agent to buy a similar product from another firm. "If that will not do," he wrote Gay, "I can strike at another course that will spoil their traf[f]ic in it." There were profits to be made in all sorts of commercial fertilizers and there was no use mourning the loss of one product when another would do as well, "There is another article of manure," he advised Gay, "which I can bring into much demand. Shall I do so & we divide the profits?"[52]

Affleck was eventually willing to act as agent for almost anything, and if he would not exact a commission in case, any form of reasonable payment was acceptable. At one time or another, he became an agent for portable steam engines, slate roofs, printing presses, books, magazines, sewing machines, stoves, sporting goods, fertilizer, livestock and life insurance.[53]

49. Affleck to D. McComb, January 23, 1851.
50. Affleck to F. D. Gay, April 10, 1851.
51. Affleck to F. D. Gay, April 23, 1851.
52. Ibid.
53. Affleck to G. C. Bryant, July 11, 1851; Affleck to H. G. Hart, July 11, 1852; Affleck to W. H. Chambers, January 19, 1854; Affleck to Colonel G. H. Peck, April 12, 1854; Affleck to James D. B. De Bow, September 20, 1855; Affleck to Robert Sears, January 9, 1851; *Almanac, 1853,* 31; Affleck to Lyall and Davidson, September 5, 1851.

If he needed some article he would often try to have the manufacturer furnish it to him so that he might demonstrate its value, and sell it on commission.[54] Failing this, he might offer to trade advertising in his publications for all or a part of the desired object and then act as an agent for its sale to his neighbors.[55] In most of the proposals he made to manufacturers, dealers, and suppliers, he emphasized his power as a writer and known agricultural consultant.

The few occasions that Affleck promoted articles of doubtful value are hardly to be considered important when compared with the great and beneficial influence he exerted in the Southwest through the various media at his disposal. He dealt with the large problems of plantation and Southern economy and the small details that made up the larger picture. He set forth his gospel with the zeal of a missionary, citing both chapter and verse.

54. Affleck to V. G. Audubon, August 19, 1853; Affleck to Colonel G. H. Peck, April 12, 1854.

55. Affleck to Lyon and Bell, September 9, 1856; Affleck to Hoard and Company, August 29, 1856; Affleck to Wallace and Lithgow, February 9, 1853.

# 2

## Excerpts from
## "The Texas Career of Thomas Affleck" (1942)

### FRED CARRINGTON COLE

*[Editor's Note: Footnotes below are original to the dissertation. Extracted from a larger work, they have been renumbered for this publication to avoid confusion and to accommodate a few additional editor's notes that add context.]*

### From "On Affleck's Writings and the Dissemination
### of Horticultural Knowledge," pages 15–20 of the dissertation

. . . A vast number of southern people were quite unaware of the value of fruits and vegetables or of the satisfaction to be derived from planned floriculture; nor did they know that the soil and climate of the South possessed many advantages for the cultivation of certain types of fruits and vegetables superior to those of any other section. There were some horticultural treatises available but none of them circulated sufficiently in the South—or for that matter anywhere—to effect much improvement.[1]

The horticultural magazines circulated as a rule only among the nurserymen and more advanced cultivators; the few southern readers viewed with suspicion the methods advocated by writers who were unfamiliar with local conditions. Southerners generally were ignorant of recent developments in dwarfing stocks, of the comparative values of certain stocks for granting, and of the different modes of propagation and pruning. Cultivation of trees, when attempted at all, was usually undertaken in the simplest and crudest

1. This is indicated by the fact that very few of the correspondents to the horticultural magazines were Southerners. The most significant and widely circulated book on the subject, A. J. Downing, *The Fruits and Fruits Trees of America* (New York, 1845), was not of much value to the Southerner, especially in selection. Gray, *History of Agriculture in the Southern United States*, 824–27.

manner.[2] There was real need for someone who could present in writing pomological developments adaptable to the South in such a manner that they
would be read without prejudice. This same need was equally evident in the
field of vegetable gardening and floriculture.[3]

Such practical matters as methods of efficient hedging had also been either unknown or neglected by most Southerners. Hedging plants were of
great importance to the planter and stock [cattle] raiser. When Affleck began to popularize the use of the Cherokee rose as the most practical hedge
for the South he was solving a long-standing problem with most ruralists.
His advocacy was not the first, but he more than any other demonstrated the
usefulness of hedging, particularly with the Cherokee rose.[4]

Thus it was not only necessary for Affleck to offer plants for sale but he
had to teach the people the names of varieties that could be grown in the
South and how to select and cultivate them. As there were no satisfactory
studies on southern horticulture, Affleck was faced with the necessity of
relying on his own experiments, upon the useful information gleaned from
authorities and publications of other sections, and upon correspondence
with those who had had practical experience in the South.[5]

Affleck was a most prolific penman. The volume of his published work
alone is impressive, but when his manuscript records are also considered,
it seems impossible that he had the time and energy to accomplish so much.
He served in the role of an embryo department of agriculture in the South,
particularly in the Southwest. He not only contributed regular columns to
the newspapers, but he also published an annual almanac and account book.
Through his personal correspondence he advised countless contemporaries,
and his advice upon almost every subject dealing with rural economy was
sought by persons from all sections.[6] The principal media for Affleck's discussions, apart from his *Almanac* and personal correspondence, were the

2. [*Southern Rural*] *Almanac*, 1851, 26–27, 38–39; ibid., 1852, 21–22.

3. New Orleans *Commercial Times*, July 9, 1846; *et seq.*; New Orleans *Weekly Picayune*, June
23, 1851, *et seq.*

4. Thomas Affleck, "The Cherokee Rose and Hedging in the South," in *De Bow's Review* V
(1848), 82–86, 175–79.

5. See for example his writings in the New Orleans *Commercial Times* during 1846 and in
the New Orleans *Weekly Picayune* during 1850–1851.

6. [*Numerous examples are given.—Ed.*] . . . The indexes to his correspondence reveal that
he had correspondents in every part of the country.

agricultural periodicals and southern newspapers. Ante-bellum magazines after 1842 carry numerous articles from his pen.[7] At various times before his removal to Texas he served as agricultural editor of the New Orleans *Commercial Times*, the New Orleans *Picayune,* and also wrote often for other papers in the Southwest. For some of his journalistic efforts he was regularly paid, but more often he was satisfied with the publicity given to his name and nursery.[8]

As a newspaper editor he prepared a weekly section running to two or three, and sometimes more columns. The type of information varied but the planter or farmer could keep abreast of the agricultural times by following Affleck's articles. There were reviews of books and periodicals which contained information useful to the South and corrected unfair or erroneous statements. . . . There were replies to queries, excerpts from personal correspondence, and accounts of his own experiments. Quite often Affleck used his column to recommend some product in which he had a pecuniary interest. At times he threatened to expose people who had treated him or some of his associates unfairly. His counsel was almost invariably good, resulted from much study, experimentation, and correspondence, and was presented in a clear, concise, readable style. It is strikingly modern in tone; for example, soil conservation by rotation, cover crops, terracing, guard drains, and commercial fertilization were proposed with much the same intent and reasoning that they are today.[9]

Although Affleck occasionally attempted to start an agricultural magazine in the South he was not successful. Two of his publications, however, circulated widely and were responsible for much improvement in the southern economy. *Affleck's Southern Rural Almanac and Plantation and Garden Calendar* was published almost continuously from 1847 to 1861, and also for 1869,[10] and his *Plantation Record and Account Book* was first issued in 1847

7. A complete bibliography of Affleck's published writings is being compiled by the author and Professor Stephenson. He published something in most of the agricultural periodicals each year. [*This bibliography, if it existed, has not been found.—Ed.*]

8. Affleck to A. M. Holbrook, . . . to A. B. Allen and Company . . . to F. D. Gay, . . . .

9. New Orleans *Commercial Times,* July 3–October 15, 1846; New Orleans *Weekly Picayune,* April 17, 1850–February 10, 1851. . . .

10. The *Almanac* was first published under the title of *Norman's Southern Agricultural Almanac for 1847* in 1846 although it was edited by Affleck. It was changed to *Affleck's Southern Rural Almanac and Plantation and Garden Calendar* for the 1851 issue. A copy of the 1851 copyright

and reprinted with revisions almost annually until 1860. The *Almanac* was the more influential of the two, for through this medium—with a circulation varying from 10,000 to 50,000[11]—Affleck presented the results of his experiments and other contributions in such a way that they appealed to numerous readers. In the absence of a significant rural periodical in the Southwest it went far toward filling the needs of the section. The principal purpose in publishing the *Almanac* was to advertise the products of his nursery. Advertisements and the sale of the *Almanacs* sometimes yielded a profit, and returns also came from advocacy of products from whose sale Affleck received commissions.[12]

### From "Looking West," pages 28–29 of the dissertation

In many ways Affleck's career in Mississippi was the most important of his life. His writings, nursery developments, scientific interests, and public services greatly benefited southern agriculture. The returns of his efforts were never sufficient, however, to provide for his family satisfactorily or to pay his debts. Although his nursery sales grossed between $8,000 and $9,000 a year after 1851 the expenses were great.[13] Labor costs took a large share of the returns.[14] It was necessary to purchase expensive fertilizers

---

is in the Affleck Papers. B. M. Norman was an influential book dealer and publisher in New Orleans, New Orleans *Commercial Times,* October 15, 1846; Harper to Affleck, December 3, 1846; *Norman's Southern Agricultural Almanac for 1848* (New Orleans, 1847), 1–3. No evidence has been located that an *Almanac* was published for 1850.

11. Affleck claimed a much larger circulation than it actually had. See frontispiece of *Almanac,* 1854, for example, where he claimed a usual circulation of 100,000.

12. Demaree, *American Agricultural Press,* lists no periodical of significance from the Southwest. There were a few short-lived magazines in the section but they were of little consequence. The *South-Western Farmer* (Raymond, Mississippi, 1842–1845); *Southern Planter* (Natchez, Mississippi, 1842), are typical.

13. This is surmised from the accounts he kept during the season 1854–1855, which seems to have been typical. Between November, 1854, and March 10, 1855, he filled orders amounting to $8,650. There were several orders made after this period but the season was calculated for the months above. Account Book, 1854–1856, 102–158. See also, ibid., 179–233; Affleck to A. B. Allen, December 11, 1854.

14. See Memorandum Book, December 25, 1854; January 23, March 3, 19, April 11, November 22, 1855, *et passim.*

because of the infertility of the soil at Ingleside.[15] Collections were not always sure and many losses resulted from the failure of plants to arrive at their destination safely.[16] Although nursery sales were greatly increased by his publications, especially the *Almanac,* there were few if any other profits from his writings after 1851.[17] Thus, it is not surprising that Affleck became more and more dissatisfied. Migration to Texas was the solution found by many people faced by similar difficulties, and Affleck's interest in this section of the United States grew during the early 1850s.

### From "Texas: A Land of Unlimited Potential, Ideal for the Entrepreneurial Spirit," pages 30–33 of the dissertation

Affleck's correspondence presents a list of the problems which he was confident would be solved by his removal to Texas. Living conditions would be improved, for the necessity of maintaining a household in keeping with his wealthy neighbors or of returning the hospitality of the older planter society of Mississippi would be eliminated. His creditors would be farther away and their demands less embarrassing. By 1854 Affleck's expenses had become ruinous, for he was obliged to buy most of the food and clothing for his family and slaves. In Texas, the slave labor problem would be diminished because of the need for workers during the entire year and the small amount necessary for their support.[18] . . .

The diversification which was possible in the new state with its cheap and plentiful lands would offer an opportunity to make the farming unit self-supporting. For example, the land adapted itself to the animals that were necessary for provisioning and working the plantation, and the cattle industry in particular could be made profitable to the plantation owner. In addition to the great staples of the older states it was said that an abundance of grapes, tobacco, rice, millet, wheat, peas, corn, indigo, and innumerable

15. Ibid., passim. Affleck to Castillo and Harispe, July 6, 1854.

16. See Account Book, 1854–1856, *passim.* Notations on the margins indicate that numerous orders were never paid for and others were settled only in part.

17. Ibid., 204–205. The accounting for the *Almanac* for 1854 which was a typical year indicates that the only profits were exchanges for advertising.

18. These factors encouraging removal are given as a result of consulting hundreds of Affleck's letters. . . .

other products could be raised in Texas. Such crops proved impracticable where land was dear and had to be used to the best advantage for cash return. It was suggested that the cost of provisioning a place in Texas would be only one eighth of that in Louisiana or Mississippi.[19]

. . .

Thomas Affleck's interest in Texas had grown with expanding nursery sales there and accompanying correspondence. A favorable notice of his *Western Farmer and Gardener Almanac,* which appeared in the Houston *Telegraph and Register,* gave him an introduction to Texans as early as 1841. As he became more closely identified with the solution of southern and southwestern agricultural problems, his reputation as a judicious and thoughtful counselor was enhanced. In fact, his writings in farm periodicals and newspapers and the widespread circulation of the *Almanac* and *Plantation Record and Account Book* soon made him as well known in Texas as in other states.[20]

In correspondence with purchases of nursery products and with his sales agents, Affleck saw the advantages to be derived from gaining a personal knowledge of Texas.[21] If he were to be as well informed as his position demanded he must get more firsthand information. During 1853 he concluded that he should shortly visit the state and make a thorough study of it with the view of getting more orders for nursery products and advertisements for his publications.

---

*[Affleck planned a series of articles on Texas as a way to increase readership there for his* Almanacs, *and these appear in its 1855 edition. He cultivated newspaper editors as a way to gain visibility for his nursery and publications. He scouted land to purchase and envisioned schemes involving land speculation, German immigrants, and railroads in efforts to make money to settle outstanding debts, but they never developed. Eventually he bought 3,400 acres in Washington County at $3 per acre, putting off creditors by writing convincingly about the prospects for profit. Beginning in 1855,*

---

19. *De Bow's Review* V (1848), 332; "Texas Lands," ibid., VII (1850), 63–64.

20. *Houston Telegraph and Register,* October 20, 1841; Matt R. Evans to Affleck, November 18,1846; Charles B. Stewart to id., February 1, 1847; October 13, 1859.

21. Affleck to O. O. Woodman, April 12, 1854; *id.* to J. W. and R. E. Talbot, March 20, 1854; *id.* to William M. Murphy, March 24, 1854; *id.* to T. S. House and Company, February 12, March 24, 1854.

*Affleck made plans to move his nursery from Mississippi to Texas, a process that was not completed until 1860, and not without the tragedy of losing his horticultural stock. Financial problems plagued him throughout; nevertheless, Affleck continued to develop schemes related to his publications, for ordering plants from American and European suppliers for his nursery, and for managing his land. Few of these schemes ever materialized, as Cole discusses on pages 34–52 of his dissertation. The Civil War erupted, as Cole notes in the following passage, and Affleck's life changed.—Ed.]*

He [Affleck] was barely settled [in Texas] in time to assume an important role as a civilian leader during the Civil War. As the master of more than one hundred slaves and the owner of many acres his experiences show the exigencies of the planter's life during the conflict. He continued to write and advise the people of the South and undoubtedly aided greatly in the section's economic improvement.

When the conflict was ended there were apparently great opportunities for those who had the foresight and initiative to take advantage of the needs of a New South, which had been helped in many ways as a result of the necessity of depending on its own resources for four years. Affleck's name was favorably known throughout the state and he was able to gain a hearing from the ideas he presented. He believed that the section must first solve its labor problems by becoming independent of the Negro, and he entered upon an enterprise for encouraging and assisting immigration from Europe to the South. He made two trips to Europe and his observations on the prospect of aid for the southern states are informative. His attempt to solve the southern labor and racial problems led him into industrial and commercial schemes of immense potentialities. Irrigation, dredging, manufacturing, shipping, food preservation, fishing, and many other enterprises were promoted by Affleck. He failed to achieve the success he had hoped for in his activities but his efforts were not futile, for he showed the way for others to follow. If all of his undertakings had been failures, however, his work would be worth study because it presents in a clearer light many of the problems of the era and the solutions that were attempted.

*[The last years of Affleck's life were not pleasant. As Cole notes, he was haunted by past business failures and economic mistakes. A scheme to secure immigrants from Europe to work the farms and ranches of Texas proved unsuccessful, and creditors demanded*

*their money back. The context for this scheme lay in the reality that, since Emancipa-*
*tion, there was an insufficient labor force for agricultural work. Affleck thought that an*
*influx of European farm workers would accomplish two things: first, it would provide*
*hard workers for the agricultural work needed to restore the local economy; secondly,*
*it would increase white populations and begin to dilute the potential political strength*
*of newly freed slaves, who were now under the influence of "Radical missionaries," as*
*Affleck wrote to the editor of the Galveston* News *in July 1867. Neither the administra-*
*tive structure created to manage the scheme nor the scheme itself materialized. Inves-*
*tors lost money, and controversy ensued, much of which played out in local newspapers*
*through published letters he and his creditors sent. Reconstruction in general and this*
*episode in particular took a toll on Affleck's energy and financial resources, and in 1867*
*and 1868 his property deteriorated because of his neglect. Crops from those years were*
*meager, as Cole notes (446–458).*

*As 1868 began, the situation began to look worse. "We are having trying times here,"*
*Affleck wrote on January 8 to William Dinn, as quoted by Cole (458–60). "In Texas we*
*suffer greatly; but nothing to the other Southern States. There, they are almost in the*
*condition of Jamaica or St. Domingo. I look daily for a war of races. Here the Negroes*
*are utterly worthless. But they manage to live—those that don't die, caught in the act*
*of killing other people's stock." Racial tensions erupted in Texas and the Ku Klux Klan*
*began to appear.*

*Affleck's debts remained, and in 1868 his plantation was sold at public auction*
*(most was purchased by a pair of local lawyers, and Affleck's wife later bought much of*
*it back). The family could only keep the homestead of two hundred acres. He tried nu-*
*merous projects to generate funds but nothing worked, and the end of Affleck's life was*
*imminent, if unexpected, as Cole reports below (464–73). —Ed.]*

---

In October, 1868, Affleck established himself in Galveston, where his druggist partner George lived, in the hope that he would be able to profit more from the sale of Cresylic compounds. His room, an empty office above the druggist's shop, was devoid of furniture and comforts of any kind. "It is not the sort of room I would like for you," he wrote to his wife, "Yet, if you can overlook the proximity of office &c., it will do for a little while, to be to-gether. I am going to get a stove in which I can make a cup of coffee."[22] He was undergoing great hardships of exposure, and his financial condition was

---

22. *Id.* [Affleck] to wife, October 15, 1868. . . .

a continual torment.[23] All the money that he could get was used to supply his family at Glenblythe.[24]

In late December, 1868, Affleck received some money from the sale of the *Almanac* advertisements, and hoped that their holiday season would be one of good cheer at Glenblythe. But even in this, he was disappointed, for the supplies did not arrive in time for Christmas.[25] The desolation at Glenblythe may have hastened the death of Affleck on December 30, 1868.

Thus ended the career of this southern leader. Only fifty-six years old at the time of his death, his busy life had been filled with sadness and disappointments. His years had never brought him personal success commensurate with his efforts. At the time of his death he had just passed another tragic climax, but new hopes for the future were appearing.

In retrospect it may be noted that Affleck's early years had prepared him to take an important role in America. After he had decided to migrate to the United States, he had carefully disciplined his interests so that he would be fitted for the New World. He had occupied his time in the study of business methods, science, and agriculture, and in extensive reading. Although he had found his extraordinary training and ability in demand in the United States, he suffered early failures and hardships. He learned through bitter experiences the trials of immigrants, both from foreign countries to America and from the eastern United States to the West. Nevertheless, this had been [a] worth-while experience, for as a writer and editor he had compared with authority the better agricultural practices of Scotland and England with those in the United States, and also those of the East with the frontier sections. After his removal to the South, he had been able to carry his comparison farther, and his advice had had soundness based upon firsthand information and thorough study.

As a nurseryman and as editor and publicist for diversified and scientific agriculture in the South, Affleck made important contributions. But the financial returns from these occupations had not been sufficient for his needs, and he sought other outlets for his energy. As a result of this search for security for himself and his family, his interests led him into nearly every phase of southern life. Indebtedness, and a natural inclination to seek

23. *Id.* to Dunbar, November 24, 1868. . . .

24. *Id.* to *Id.*, December 10, 1868. . . .

25. *Id.* to Dunbar, December 16, 1868; *id.* to A. Whiting, December 21, 1868.

greener pastures, had caused him to remove to Texas, where he had barely settled in time to assume an important role as a civilian leader during the Civil War. His advice and leadership were invaluable to the people who had been forced at last to become self-sufficient, but his personal fortunes had not been improved.

When the conflict ended, apparently there were great opportunities for those persons with the foresight and initiative necessary to take advantage of the needs of a New South, which had been helped in many ways as a result of having had to depend upon its own resources for four years. Few men had the energy and courage to put their panaceas to a test. Believing that the section first had to solve its labor problems by becoming independent of the Negro, Affleck entered upon his enterprise for encouraging and assisting immigration from Europe to the South. His unsuccessful attempt to solve southern labor problems led him into industrial and commercial schemes of immense potentialities, and finally, by chance, to the publicizing and sale of creosote compounds.

Despite his failings, Affleck's work was of immeasurable value to the country. Many of the programs, plans, and enterprises that he advocated were hastened by his efforts and have since been realized. Some of them, such as the appointment of an immigration commissioner, were to follow soon after his death; because of his work, interest in immigration of white people to the South increased greatly during his life time.[26] Diversified farming and improved implements, about which he had written so much, were soon to be in evidence on almost every farm in Texas;[27] regular transportation to Europe was shortly provided,[28] and better harbors for Texas were to come somewhat later; his work in the interest of the livestock industry must be given much credit.[29] Irrigation and manufacturing of various sorts were forecast by Affleck, who had encouraged their development. If all of

26. See, for example, Galveston *Daily News,* March 9, 10, May 17, 18, 19, . . . 1867; Houston *Weekly Telegraph,* July 23, October 22, 1868; Houston *Daily Telegraph,* June 4, July 1, . . .

27. Galveston *Daily News,* May 10, 1867, *et seq.,* for reports on the sales of implements.

28. Ibid., July 9, 1867; Houston *Weekly Telegraph,* November 26, 1868.

29. See Galveston *Daily News,* July 10, 1867; July 3, September 1, 1868; Houston *Weekly Telegraph,* October 1, 1868; *Flake's Galveston Daily Bulletin,* July 1, 1869, for notices concerning the movement for preserving beef in Texas and for the shipping of cattle out of the state. His most important contribution in this regard was, of course, his introduction of Cresylic acid.

his efforts had been futile, however, his work would still have been worthy of study because it presents in a clearer light the history of his time.

In order to understand Affleck's great influence upon his contemporaries, his personal characteristics and the power of his pen must be emphasized. He gained the immediate attention and usually the confidence of those with whom he was associated. Handsome as a youth, he had become more distinguished in appearance as he had grown older. His dignified bearing and naturally serious countenance, made more prominent by a full beard, marked him out among men.[30] An old lady who remembers him when she was a very small girl retains a vivid picture of his bearing and features. She has written in her memoirs that "he was rather stern in appearance, but kind in heart. All of us stood greatly in awe of him."[31] His keen intelligence had been more impressive as a result of his remarkable command of information upon the most varied subjects. He had always been at ease when conversing, and he had been equally apt at making public addresses. His style as a writer had been lucid and readable, although not polished. For the rural population, to whom it was primarily directed, it had been ideal.

Personal faults loom large in the career of Affleck. He had never been able to undergo periods of hardship without trying to find an easy solution to his problems. He does not appear to have been personally avaricious, but he had sought to make money to pay his debts and provide for his family in ways that were sometimes of questionable ethics. As a result, his conscience was at times respited only by his facile logic and pen. From the vantage point of the present it appears unwise for him to have undertaken many of the enterprises and promotions which drained so much of his energy. Immense sums of money would have been necessary for the successful culmination of some of his schemes. Although he usually had been promised more support than was forthcoming, in many instances he did not show good judgment nor take proper precautions. Nevertheless, he made people aware of solutions to many of their problems, and he presented workable plans for others with the means to fulfill them; he exposed faulty theories, if he did not always effect changes.

30. There is a picture of Thomas Affleck made in the 1850s in the home of Mr. T. D. Affleck at Galveston, Texas.

31. Foules, "My Memories."

Most of what Affleck undertook was basically sound. As a visionary and experimenter himself, he considered the advanced ideas of others without prejudice and often was able to present scientific and mechanical discoveries and new developments in commerce and industry. The Houston *Telegraph* soon after his death gave a key to his place in history in the following notice:

"He was not only a good, but in our opinion a great man, and at least fifty years in advance of the age in which he lived. As a man of science, he occupied a proud position, but not one equal to the greatness of his conceptions. He did much as a writer, for the orchard, the garden, and all branches of agriculture, and the record he made of his greatness will be better appreciated now that he has left us. Because he was ahead of his age, many thought him visionary, but the day will come when men will say that a monument should be erected to his memory. Had he possessed the means, he would have developed his plans, formed in his great mind, that would have placed him far up among the great benefactors; but the want of the means crushed down his aspirations, and others, not half so worthy, will reap glory from his conceptions in the future. We regret to put this on record—but who, that has read of Fulton, Watts, and a hose of public benefactors, will dare dispute? THOMAS AFFLECK was a benefactor, and the day will come when the world will honor his memory."[32]

32. Houston *Telegraph,* quoted in a printed letter sent out by James Buchan and Company, January 11, 1869.

# II
## SELECTED WRITINGS
### 1840–1869

# 3

## From *Western Farmer and Gardener*

American agricultural newspapers and journals were published first in the New England states, but as people moved west, so did newspapers and journals. One of the first and most influential was the *Western Farmer and Gardener,* published first in Cincinnati from 1839 to 1845, when it ceased publication because of the failing health of its primary editor, Edward J. Hooper. In 1846, the *Western Farmer and Gardener* was combined with the *Indiana Farmer & Gardener* of Indianapolis and continued publication as the *Western Farmer & Gardener* under the joint editorship of Hooper in Cincinnati and the Reverend Henry Ward Beecher in Indianapolis.[1] The paper's focus shifted to the Indianapolis region, and there were few articles related to locations, crops, or climates outside that region; the paper soon lost its southern readers. Beecher became sole editor in 1847, but issues from that year have not been found, and it is not known whether the publication continued after 1847.

The *Western Farmer* was popular among landowners and community leaders in its region, and it was well known for its articles on horticulture and livestock, two subjects about which Affleck wrote extensively.[2] While it was published in Cincinnati, the *Western Farmer* had several editors: Hooper, Thomas Affleck, Charles Foster, and Charles W. Elliott. Affleck was named associate editor in 1840 and became editor in 1841, serving in that capacity until he moved to Mississippi in 1842.

As was common during this period, agricultural newspapers (or "papers" as they were then known) such as the *Western Farmer* contained original

1. Henry Ward Beecher (1813–87) was a prominent nineteenth-century clergyman, social reformer, and abolitionist. One might expect he would publish editorials and articles that reflected his political and religious leanings. Judging from available 1846 issues, there was little—if any—such content.

2. For more discussion on the *Western Farmer* and other such agricultural newspapers, see Demaree, *The American Agricultural Press,* 16.

staff reporting, contributions from subscribers, and articles reprinted from other publications of the period, such as the *New England Farmer* (Boston, 1822–46); the *Cultivator* (Albany, New York, 1834–97); the *Maine Cultivator* (Hallowell, Maine, 1839–50); and the *Agriculturist* (Nashville, Tennessee, 1840–45). The *Western Farmer* sent Affleck on extended trips, first through Kentucky and other neighboring states and then to Mississippi and Louisiana, with the purpose of making first-hand observations about agricultural conditions (both crops and livestock) and activities (including fairs, conventions, and societies) in these areas. His reports were then published.[3]

Representing the *Western Farmer,* Affleck doubtless sought to enlist new subscribers, since there were no regional sources of agricultural information readily available for residents of the states that he toured. In reading his reports, one concludes that Affleck was personable and congenial. Judging from names he mentions, we learn that he quickly became acquainted with community leaders, landowners, and people of influence and social prominence in the communities he visited. An entrepreneur, Affleck had taken some of his own livestock with him to show and sell at the agricultural fairs he visited as a way of promoting his credibility (and that of the publication he represented), advancing current practices of animal husbandry, and making money. He also met his future wife in the Natchez area while on his visit there. He left Cincinnati and the *Western Farmer* in 1842, moving south to marry her and manage her plantations, which, he soon found, were in financial and agricultural disarray.

Affleck likely realized that he could use his expertise in animal husbandry and horticulture, his experience as a writer and editor for the *Western Farmer* and editing its 1842 *Almanac,* and the contacts gained from his travels to generate new business opportunities in the Natchez area.[4] While there were numerous American agricultural newspapers in the 1840s, most were short-lived and limited in their geographical coverage to regional au-

---

3. These first-hand observations resemble in many ways those made over a decade later by newspaper journalist Frederick Law Olmsted on his journeys through the South.

4. This advertisement appeared in the *Ohio Statesman,* Columbus, in September 9, 1841: "Excellent Publication. We have received from the author, '*The Western Farmer and Gardener*'s Almanac for 1842' by THOMAS AFFLECK, editor of the Western Farmer and Gardener, Cincinnati, Ohio. This Almanac contains 150 pages, and [is] closely filled with the most invaluable matter for the farmer and gardener—and who is not one or the other? We do not know when we have seen anything that we more highly approved of than this work. Buy it!"

diences. Curiously, few were published in the agricultural South, and at the time none covered Mississippi and Louisiana.[5]

As markets changed over time, publications disappeared or consolidated to continue publication, and places of publication changed.[6] Many had similar names (often with nouns such as *cultivator* or *farmer* or adjectives such as *agricultural* and *horticultural* in their titles), making it difficult today to track evolutionary chronologies.[7] Complete runs of such publications are difficult to find today because production values among these "papers" were modest, if that. They were cheaply printed, and most subscribers eventually discarded copies after reading them. Nevertheless, editors urged subscribers to keep issues for future reference, and publishers often bound back issues and returned them to subscribers. In many cases, today's examples exist because they were bound as books.[8] For most of the nineteenth century, there were few institutions or libraries that collected works on this subject; hence most extant examples of these publications are single issues or bound volumes of a year's collection of issues, and several are now available electronically.

The *Western Farmer* may well have been one of the most important agricultural papers of the mid-nineteenth century because of the diversity and quality of its content, and its wide regional influence. In addition, the *Western Farmer* was unique in that its editors extracted the "largest and best parts" of the publication and re-published the collected articles in two hard-

5. Stephen Conrad Stuntz, comp., and Emma B. Hawks, ed., *List of the Agricultural Periodicals of the United States and Canada Published During the Century July 1810 to July 1910*, Miscellaneous Publication № 398, United States Department of Agriculture (Washington, D.C.: Department of Agriculture, 1941). According to this source, the *Tennessee Farmer* appeared in 1836 (Jonesboro); the *Southern Agriculturalist and Register of Rural Affairs* (Charleston, South Carolina, 1828–39), later the *Southern Agriculturalist, New Series* (Charleston, 1841–46); the *Tennessee Farmer and Horticulturist* (Nashville, 1846–47); the *American Cotton Planter* (Montgomery, Alabama, 1853–61); the *Mississippi Planter and Mechanic Devoted to Agriculture, Horticulture, and the Mechanic Arts* (Jackson, 1857–58).

6. See Demaree, *The American Agricultural Press;* and Stuntz and Hawks, *List of Agricultural Periodicals*. For discussion of agricultural journals and their content, see Lake Douglas, "To Improve the Soil and the Mind: Using Context and Content of Nineteenth-Century Agricultural Literature for Environmental Research," *Landscape Journal* 25, no. 1 (2006): 67–79.

7. Stuntz and Hawks, *List of Agricultural Periodicals*, is perhaps the only source for this information.

8. The *Horticulturist* (Andrew Jackson Downing was its editor from 1846 to 1852) did this for its subscribers, thereby ensuring this journal's place in many American libraries.

cover editions (at 1,168 pages each) as *The Western Farmer and Gardener: Devoted to Agriculture, Horticulture, Gardening, the Flower Garden, Cattle Raising, Etc.*, first in 1848 and again in 1850 (used here). The fact that there were two hardbound editions of these collected works suggests that the material was useful and well received by the public of the late 1840s and early 1850s in the rapidly expanding West and Southwest, the present-day states of Ohio, Indiana, Illinois, Kentucky, Tennessee, Missouri, Alabama, Mississippi, Louisiana, and Arkansas. Certainly Affleck's contributions from first-hand observations "in the field" describing agricultural life in Mississippi and Louisiana increased readership in that region, added to the *Western Farmer's* constituency, and stimulated interest in agricultural institutions such as local agricultural societies and perhaps even regional agricultural publications, such as the *Mississippi Planter & Mechanic*, from 1857.[9]

Content in these agricultural publications covers a broad range of topics related to agriculture, animal husbandry, and domestic life; some even contained "women's departments" that covered household hints and recipes. There were also magazines devoted wholly to women's issues, such as *Godey's Lady's Book*, published in Philadelphia between 1830 and 1878. Prior to the Civil War, *Godey's* was the most popular periodical in America, with circulation in 1860 reaching 150,000. Its contents included instructions regarding sewing and needlework, poetry, short stories, and articles (mainly by women), piano music, and, notably, colored fashion plates. Curiously, however, there were few if any articles related to gardening or domestic horticulture.

Agricultural papers were the stages on which agricultural issues were presented and the forums through which agricultural reformers and their platforms became widely known. Editors were influential figures, and their publications advanced agricultural education, efficient farm management, new growing techniques, technological advancements of equipment, and "fluffing," when manufacturers paid editors for prominent product placement and "unbiased" editorial discussion of new products.[10]

Politics in general and the politics of slavery in particular were hardly discussed, even though some agricultural reformers by the late 1840s and early 1850s were beginning to realize that the single-crop economy prevalent

---

9. To date, only Volume I, numbers 10–11 (October–November, 1857) have surfaced.

10. Some editors scrupulously resisted such practices and would only discuss products they had tried and felt met their manufacturers' claims. Affleck often mentioned new products, but it is not clear whether he had personally tried everything he discussed.

in much of the South, produced through a slave-based economy, was both agriculturally inefficient and financially unsustainable. An editorial in the *Western Farmer* (May 1842) addressed to "the farming community" begins: "It is part of our compact with our subscribers not to talk politics, and we believe we will not overstep our bounds in the following observations. The pressure of our times is such, that we feel compelled to notice it, and our sole object is to state a few circumstances we feel worthy [of] the attention of the farmers of the country." The editorial then notes how much the country depends upon farmers and calls for the establishment of agricultural societies, for the careful attention to the business of being a farmer, and for "improvement, moral and intellectual." Further, the editorial encouraged farmers to "increase the amount of their crops and improve their breeds of cattle," thereby improving their own social condition in order to add to the "wealth of the country and elevate her standing with the world." This was support for agricultural reform that altogether sidestepped slavery—the agricultural elephant in the country's parlor.

Journal articles are as often signed as not, and when reprinted from other sources, credit is sometimes given, sometimes not. In the *Western Farmer,* authorship appears as a name or as initials, and dates and sources of original publication are inconsistently given. We may assume that articles in the bound volume of collected works are given in the chronological order in which they originally appeared; however, in this work, articles are grouped by general subject matter rather than the order in which they appear in the collected volume, with notations given of where they appear in the bound volume (1850 edition) used here.

This "Notice" appears as front matter in the *Western Farmer:*

### NOTICE.

*The "Western Farmer and Gardener" was originally published periodically, running through a series of five volumes, from September, 1839, to August, 1845, inclusive, under the management and editorship of E. J. Hooper, Thomas Affleck, Charles Foster, and Charles W. Elliot, Esqrs. The contents of this volume are made up from the largest and best part of those five; such matter being omitted as had particular references to local affairs, transpiring at the time of publication, and seeming to possess but little interest or value to the future reader.*

*The "Western Farmer and Gardener" is a practical book; the editors, and other able writers for the work, having had long experience in the different branches of agriculture, Horticulture, Gardening, Cattle Raising, &c., &c., upon which they treat; and the selected articles being taken from the best works and journals on those subjects, all having especial reference to the climate, soil, &c., of the South and West, and Northwest.*[11]

In this compilation of work from the *Western Farmer,* Affleck's first article is "Planting Fruit Trees," followed by some forty more, the last one appearing in 1843 after Affleck had relocated to Washington County, Mississippi. As is evident from the titles of his contributions, Affleck wrote on a variety of subjects, including general discussions of gardens, flowers, and fruit trees; beekeeping (a particular interest, and one that appeared in 1841 as a monograph published in Cincinnati); creating agricultural societies; and efforts to improve livestock, notably hogs (another particular interest). Affleck also wrote extended pieces on conditions, concerns, and issues based on first-hand observations in Kentucky, Mississippi, and Louisiana, a narrative device employed by many during the nineteenth century (notably Frederick Law Olmsted in his journeys through the South in the 1850s). In these accounts, we see in Affleck's words—as in those of Olmsted a decade later—an acknowledgment and appreciation of the environment together with the recognition of how the landscape might be organized for both profit and pleasure.

Throughout his career, Affleck's writings are "in the plainest possible manner" (words he used in an early article to describe what he valued in contributions from correspondents), increasing their accessibility. The selections given, representing his early career as a journalist, are more about horticultural topics than animal husbandry. These subjects, together with others discussed in the *Western Farmer and Gardener,* reflect the interests of the paper's readers and offer a window through which we can view mid-nineteenth-century American horticultural and gardening concerns.

All footnotes to the text are my own annotations, which are intended to provide additional information or context. The page numbers provided in the first note in each article reflect the pages in the *Western Farmer and Gardener* compilation (1850) from which the article is taken.

—Ed.

11. *Western Farmer,* 6.

---

## Let Us Have a Horticultural Society![12]

Little towards improvement in gardening can be done without one. Witness the astonishing results of the establishment of societies in the East and in the old country. To their influence we are indebted for the many new and delicious fruits, the improved vegetables, and the beautiful shrubs and flowers which the enterprise of a few individuals has introduced amongst us—sparingly, it is true, yet enough to show us what *could* be done by the concentrated efforts, of even *the scanty number* among us who take an interest in such things. The societies established in Boston and Philadelphia have been the means of supplying the markets of those cities with *an abundance* of fruits and vegetables, such as *we* must be satisfied to hear their citizens boast of, but cannot, as yet, partake of ourselves. We are fully persuaded that the soil and climate of the valley of the Ohio, and more particularly around our own city [Cincinnati], are much better adapted to the growth of fine fruits and flowers, than in many parts of the East where they are cultivated to a great extent. We see that the Massachusetts Horticultural Society offers a premium for the discovery of some means of checking the ravages of *a slug*, which there infests the rose-bush to such an extent as "to rob the fairest flower in creation of its beauty." We never heard of nor saw any such insects here: nor do we believe, that with one tenth part of the care bestowed *there* on their cultivation, would we have to complain of *blight* in the *pear,*

12. From *Western Farmer and Gardener*, 237.

*yellows* in the *peach,* nor of the ravages of the *curculio* or the *borer,* nor of many other annoyances, of which the gardener has too often cause to complain. But alas! We have greater enemies still to struggle with—neglect and indifference. We have apples, pears, and peaches, vegetables and flowers, *such as they are;* some few are good, but by far the greater part unworthy of garden room. All this can be remedied. Let the gardeners bestir themselves—excite each other to take more interest in their profession, and, by the establishment of a Horticultural Society, become acquainted with each other, and with the progress in improvement they have individually made, and with that made elsewhere—by the exhibition of fruits, vegetables and flowers, and giving premiums for the best specimens of each, "stimulate and reward industry and enterprise, however humble their condition, and strive, by concentrated and persevering efforts, to improve the condition of a district, of a country or a state—inspire public confidence, obtain public blessing."

To all, but more especially to our fellow gardeners, to amateurs, to those of our citizens who love good fruit—who admire and wish to possess a beautiful flower—to all such we would say, think of this matter, view it in its proper light, and when we call on you to attend a public meeting for the purpose of organizing such a society, and which we hope ere many months to do, come prepared to join us, heart, hand, and purse, in the good cause. We will return to this matter in our next, and in the mean time would say, that our columns are open to discussion on the subject.

T. A.

# GROWING FRUIT

The first books published for American audiences on agricultural matters had lengthy chapters with general information on growing techniques for fruits and vegetables, and by the end of the 1840s, books had appeared by leading agricultural experts on specific subject areas and plants such as fruit trees, the kitchen garden, and roses. Among the most notable examples are Bernard McMahon's *The American Gardener's Calendar: Adapted to the Climates and Seasons of the United States* (1806); Thomas G. Fessenden's *The New American Gardener* (1828); Thomas Bridgeman's *The American Gardener's Assistant* (1835); Robert Manning's *Book of Fruits: Being a Descriptive Catalog of the Most Desirable Varieties of Pear, Apple, Plum, Peach, and Cherry for New-*

*England Culture* (1838); William Kenrick's *The New American Orchardist* (1844); Robert Buist's *American Flower Garden Directory: Containing Practical Directions for the Culture of Plants* (1845); and Andrew Jackson Downing's *The Fruits and Fruit Trees of America* (1847). All went into multiple editions throughout the nineteenth century.

Agricultural newspapers, cheaper and more widely available than books, contained similar content, with much space devoted to advice on varieties of fruits and vegetables, cultivation techniques, and the management of insect pests and diseases. These papers often repeated general agricultural practices, but editors also published articles on regional topics submitted by subscribers or lifted from other agricultural papers. They also did not hesitate to ask for information from their readers in efforts to gather practical information "from the field," as in an exchange about peaches in which Affleck notes a curious phenomenon about peach culture and seeks responses from his readers. A reply was received and subsequently published, characteristic of the camaraderie and exchange among editors and subscribers. In the following discussions of watermelon, grapes, and wine-making, note how Affleck refers to personal experiences and observations.

—Ed.

## Planting Fruit Trees [13]

The season is now at hand when it is necessary for those who intend planting out fruit trees, either in the orchard or the garden, to exert themselves. It is much better to do so in October or November, than to put it off til spring. Not only is it better for the trees, as they are thus enabled to establish themselves in their new situation before the spring opens and they bud out—the fall and winter rains settling the earth firmly about the roots, which are by no means idle during the long winter months—but the farmer has more leisure now than in the hurry of spring work, and can take pains to throw out a hole for the reception of each tree, *at least* four feet across, loosening up the soil at bottom, and adding some fresh light mould from the woods or roadsides, that the young roots may have a good start; *staking* the tree firmly and piling a few large stones, inverted sods or long litter round each, to keep all solid and snug.[14] All this is necessary where it is expected

13. Ibid., 234.
14. *Mould* and *litter* were terms used to describe decayed vegetative matter.

that the trees will grow and thrive. At this season of the year, too, the *nurseries* are well stocked; the handsome, healthy-looking plants have not been culled out, leaving generally but a poor choice for those who have neglected planting till spring.

T. A.

## Pears [15]

There is a subject connected with the cultivation of the pear, upon which we should gladly be informed. Pear stocks being much more inaccessible to our nurserymen, than apple, they are in the habit of working the former on the latter, and do so the more readily, as it has been found here that the Seckel, and some of the other smaller pears, are improved by being grafted on the apple.[16] We ate a Seckel pear a few days ago, of a most unusual size and flavor, taken from such a tree. Will this hold good with the large melting juicy pears? and, what is of still more importance, how does it affect the *durability* of the tree? We find it generally allowed, that fruit of any kind is affected to a certain extent by the nature of the stock used. Miller says, decidedly, that the common Crab stock causes apples to be firmer, to keep longer, and to have a sharper flavor; and he is equally confident that if the breaking pears be grafted on quince stocks the fruit is rendered gritty or stoney, while melting pears are much improved by such stocks. Therein he agrees with what we have stated as the result with the Seckel pear, which on a pear stock, is small, though free from *grit,* which we should expect to find if on a quince or thorn stock. This, however, does not agree with Lord

15. From *Western Farmer and Gardener*, 239.

16. When grown from seeds, fruit trees such as apples and pears do not necessarily breed true to the parent. European immigrants commonly brought seeds from their homelands, planting them in the farmsteads on which they settled, and therefore many new varieties of fruit trees emerged in America throughout the nineteenth century. This explains much about the evolution of horticulture in America: the early interest among many for joining "pomological" and agricultural societies; the common occurrence in agricultural journals such as the *Western Farmer* of articles devoted to fruit trees; the early attention from horticultural writers to the subject; and the subsequent appearance of horticultural manuals devoted to the cultivation of fruit trees, which were among the first horticultural books to appear in America. The Seckel pear, likely a wild seedling, is thought to have been introduced in the early nineteenth century by a Pennsylvania farmer for whom it is named. A small, roundish pear with firm flesh, it is better suited to cooking and preserving than to eating off the tree. It is still grown and commercially available today.

Bacon's doctrine, that "the [s]cion overruleth the graft quite, the stock being passive only," which, as a general proposition, is in a great measure true.[17] The distinct characteristics of the *engrafted* fruit remaining unchanged, although the qualities of the fruit itself are partially affected. Will some of our eastern friends, Messrs. Hovey, Manning, Ives, Downing, &c. (our acquaintance with all of whom, though not *personal,* is by no means of yesterday,) who have been enabled to put the experiment to the test of time, enlighten us?[18] I have found, in my own experiments, that where the quince or thorn stock is used, the graft lives and thrives a few years, perhaps bears once or twice, but soon shows signs of decay, and in five or six years dies. The late Mr. Knight observes the same thing, and gives as his opinion, that "a stock of a species or genus different from that of the fruit to be grafted upon it, can rarely be used with advantage, unless where the object of the planter is to restrain or debilitate." In grafting the pear upon the apple root, we have rarely (though in a few solitary instances we *have,*) found that a union took place between the root or graft—yet the pear grows, being evidently supported by an absorption of *moisture* at least, from the root, until it throws out roots of its own. We have repeated enquiries made of us by individuals who wish to plant out pear orchards, as to the best varieties to be used, and particularly of the very early kinds, and in our answers were guided principally by the published experiences of the gentlemen named above; but owning to the limited stocks and assortments of choice pears here, as yet, few of the fine varieties there spoken of can be obtained—we know of only a few hundred fit for market. Will the gentlemen we named oblige us by forwarding to the office of the Farmer, a priced catalogue of all their choice fruits, and particularly those new kinds which have fruited with them, checking off those

17. The reference here is to English philosopher, statesman, and essayist Sir Francis Bacon (1561–1626), often considered the father of modern science.

18. Affleck knew of Hovey, Manning, Ives, and Downing by their reputations as leading American horticulturists and horticultural writers of the late 1840s. Charles Mason Hovey (1810–87), one of America's leading nurserymen, horticulturists, and horticultural writers, lived in Cambridge, Massachusetts, and was particularly interested in fruits and ornamental plants, notably camellias and chrysanthemums. In 1836 he introduced the "Hovey Strawberry," considered to be the start of strawberry cultivation in America. Hovey was editor of the influential *Magazine of Horticulture,* modeled after Loudon's *Gardener's Magazine,* which had an uninterrupted run from 1835 to 1868, the longest of any American horticultural journal to date. Also, Hovey wrote *Fruits of America,* issued in two volumes from 1852 to 1856. With more than a hundred colored plates, the book's purpose was to describe the "choicest varieties cultivated in the United States."

they have not already spoken of? By doing so they will benefit the cause we advocate, and enable us, probably, to send them some orders.

T. A.

———————◆———————

### Peaches.—Information Wanted [19]

I find it to be the prevailing opinion amongst many old farmers in Virginia and Kentucky, that where a peach-kernel, the produce of an original or unbudded tree, is planted, the product is a peach closely resembling the parent fruit. On the other hand, where stocks are raised from the kernels of indifferent peaches, and upon these are budded the finest varieties, the kernels from the fruit of such trees produce trees that sport very much, the peaches being almost always small and worthless. The inference from this is that the stock so far overrules the bud, as to cause the produce to partake almost entirely of its own character. One intelligent and experienced fruit-grower of this city, with whom I have recently conversed on the subject, states that the results of his own experience would induce him to believe that such is the case; yet he cannot be satisfied that it is so. He thinks, with me, that if a kernel was planted that grew on a tree, the stock of which was the growth of the kernel of a cling-stone peach, one which was afterwards budded a freestone, that the product would be freestone peaches; and that the rule would hold good, whether the stock grew from the kernel of an indifferent, or of a fine peach, the [s]cion or bud would still overrule it. The truth is, the peach, like every other cultivated fruit, sports, so that no reliance can be placed on a seedling. We consider it an ill-judged waste to time and ground, to cultivate any but trees on which have been budded choice and well known fruit. Still we are anxious to have information on the subject, from those who have been able to put it to the test of careful and judiciously conducted experiment. Our pages are open to the discussion of this and every other useful subject connected with farming and gardening.

T. A.

*[A reply was received and subsequently published that answers Affleck's questions. —Ed.]*

19. From *Western Farmer and Gardener*, 259.

## Cultivation of the Water Melon [20]

There is no plant more tender or difficult of cultivation than the Water Melon. It is of a different genus from the Musk Melon, which is, in comparison, hardy and certain in its crop. The latter may be grown in great perfection, in the same manner, as we will now direct for the cultivation of the former; with this difference, that it may be planted a few days earlier; will bear transplanting from a hot-bed; and need not be over from 7 to 9 feet, from hill to hill, according to the kind.

The best soil for the Water Melon is a piece of meadow, or old sod-land, ploughed the August preceding, as deep as possible; left to lie in the rough all winter; early in the spring cross ploughed; and if not pretty rich, had best have a good coat of *old well-rotted* stable-yard or chip manure, scattered over it, previous to the second ploughing. This is better than putting the manure in the hill; as on examination, it will be found that the roots of this plant spread over the whole space, from hill to hill, and moreover are always to be found within three or four inches of the surface. About the third week of April—I speak of the vicinity of Cincinnati—let the ground be marked out, but running light furrows with the plough, twelve feet apart, crossed by others at the same distance. Where these intersect, form a flat hill of 15 or 18 inches in diameter, and not over three inches in height. Immediately, if the weather is mild and favorable, plant 6 or 8 good sound seeds at one side of the hill. As soon as these appear above the ground, or on that day week after they were planted, put as many more of the same variety at the other side— and it is even good practice, to make a third planting one week after the second. You thus secure a sufficiency of good, stout plants in each hill. When the plants are yet in the cotyledon or seed leaf, go over each hill, and with the finger gently withdraw the earth from about each plant, nearly to where the rootlets appear. This will in a measure prevent the cut-worm's attack; will harden the stem, and prevent its damping off, if a cold wet spell come on. Leave at least 8 or 10 good, stout plants in each hill, thinning them out gradually, as they seem to require room—until, when they begin to vine out, all may be removed but two. In the meantime, the ground must be kept clear of weeds by the use of the plough or cultivator, and hoe. When the plants have vined out about a foot, go over them for the last time; leveling off the ground, and forming a neat, flat hill, of two feet and a half round the plants, but taking care *not to draw any earth up to the stems.* After this they require

20. Ibid., 349–50.

no farther attention. A *few* weeds scattered about, of a trailing growth, will be of advantage rather than otherwise; for if there is nothing for the plants to attach themselves to, by their tendrils, the first gust that comes, will blow them about in every direction, to their certain injury—they ought never to be disturbed or *displaced.* Care should be taken to procure sound seed, of a good sort. The best and largest we have met with were the Black Spanish, the Long Island, the Ice Rind, and Goodwin's Imperial.

The cut-worm is the great enemy to our water melons, sweet potatoes, and early corn. The best *cure* is fall and winter ploughing—but where this has been neglected, much of the picking by hand, I have been assured by a gentlemen of experience, may be rendered unnecessary, by scattering a little salt round the margin of each hill—that the worm in its night journeys in search of food, will *never cross the line thus drawn.*

It has often been attempted to grow the water melon in a hot bed, and transplant, but never to our knowledge, with success. It might however be done, by planting in small pots, taking care to harden the plants before setting them out, and carefully turning the ball of each, enclosing the roots, into the hill.

T. A.

———————◆———————

## The Grape[21]

Here we have a sketch of the mode of pruning and training the Vine, as practiced in the vineyards of this vicinity—not carried out to its full extent, which the space we can afford will not allow it; still it will be sufficient to explain the system to those who have never seen it practiced. The vines are represented as in full leaf and bearing, and after the young wood of that season had partially made its growth.

Now is the time to prepare the cuttings, and to make new plantations of the vine.

The first things to be considered are the proper location of the vineyard, and the preparation of the ground.

We have made a practice of visiting every vineyard within our reach, and have remarked that those situated on the north, or north-east sides of a hill,

21. Ibid., 333–34.

have borne the most certain and abundant crops, and we believed, as well ripened. The warm, unseasonable weather that we almost invariably have in February and March, starts the flow of sap in those vines on a southern exposure, and throws them too soon into leaf and blossom, which are almost always destroyed by the frosts in the end of April or the first few days of May. So it is, too, with the peach and the early apples, etc. There is no danger of the fruit not ripening, in a northern exposure—we have rarely the slightest lack of sun and heat at that season! The hill tops are preferred, for various reasons—not the least satisfactory of which is, that it is the most profitable use to which they can be put. A moderately rich loam, on a clay subsoil, with a substratum of limestone, has always seemed to us the best adapted to the production of *sure crops of fruit.* In rich bottom land the vine itself grows with much greater rapidity, but the fruit is apt to mildew. Where practicable, the ground ought to be trenched with the spade, at least two spits deep—thus securing the rich surface soil at a depth which secures it from washing. If this cannot be done, and the ground was not ploughed to a sufficient depth last fall, let it now be *double ploughed;* running two ploughs in the same furrow, one behind the other—and let this be done as soon as possible.

The next thing to be considered is to procure good, well-ripened, properly prepared cuttings, and *true to their kind.* On this depends the main success of the vineyard. They are formed of the wood of last season's growth, cut square immediately above and below a joint; and from fifteen to eighteen

ˏ inches long. Let them be prepared before the sap begins to flow, and buried in a cool cellar. If they have to be carried any distance, they ought to be carefully packed in moss. In the meantime let the holes be dug for their reception, eighteen or twenty inches in diameter, and about the same in depth. About the middle of April, after soaking the cuttings twenty-four hours in a tub of rain water, plant them out, two in a hole at opposite sides, and at such a depth as to allow two buds only to appear above ground—fill up the hole with rich, light mould; if somewhat sandy, so much the better. Tread the mould firmly at the *base* of the cutting, and moderately so along its whole length. If planted along a trellis, about five feet from plant to plant, is a good distance. If in a continued and regular vineyard, the rows may be six feet apart, and the plants three feet from each other in the row; this allows of the strawberry being cultivated between the vines, which not only keep the soil from washing on the hill side, but will more than pay all the expense of cultivation. A row of sugar beet or of potatoes may be grown between each row, the first season.

We will return to the after treatment in a subsequent number.

T. A.

————◆————

### Wine-Making [22]

Cincinnati, O., 9th March, 1842.

*Isaac Dunbar, Esq. of Washington, Mississippi:*

MY DEAR SIR—I feel a great desire to give you what information I can gather, relative to the *modus operandi* of the *vignerons* of this vicinity in wine-making; but fear that I shall not be able to be sufficiently minute to aid you much. But I shall do my best. It unfortunately happens that those few who do possess any knowledge or experience in such matters, are either such men as are not in the habit of writing, or are not public-spirited enough and sufficiently obliging to take upon themselves the trouble. If it had happily been otherwise, you, and others persevering and public-spirited as you, would not have had to struggle on through years of unsuccessful experiments to arrive at a good result; but would at once have been put on the proper track, and have been enabled more quickly to arrive at success in your efforts.

---

22. Affleck is writing to his future father-in-law. Essay from *Western Farmer and Gardener*, 560–62.

I shall leave you to refer to your books for everything connected with the comparative qualities of the *must*[23] and for all such particulars as you will be likely to find there, and confine myself to a few simple items, learnt from my own observation, or from one of our most successful *vignerons*—Mr. John E. Mottier.[24] As I mentioned to you in conversation, some short time ago, the varieties of the grape cultivated here, for wine-making, are the Catawba, Isabella and Cape or Schuykil Muscatel.[25] Another variety has been recently much lauded, and I think deservedly—not however as a wine, but as a table-grape. I allude to the variety called by Mr. Longworth the *cigarbox*.\* Two cuttings were given to Mr. Longworth some years ago—left at his house in a cigar box, by someone, without any particulars accompanying them. They were engrafted on strong stocks, and have since become large vines, as have also those propagated from them. I spoke of the variety to yourself and other gentlemen in Mississippi; and I bore to Mr. Longworth the polite request of the following gentlemen to be furnished with even a single cutting or two:—yourself, and Wm. St. John Elliot, J. F. H. Claiborne, Wm. J.

23. *Must:* from the Latin *vinum mustum*, "young wine," the first step in wine-making, it is the freshly pressed juice of the grape with the fruit's skins, seeds, and stems.

24. John E. Mottier (1801–88), the "wine king of Cedar Avenue," lived in the Cincinnati area and was well known in the region for his "Delaware Vineyard." Affleck had lived in Cincinnati, and obviously they knew each other through a shared interest in agricultural activities. Born in Switzerland of French descent, Mottier came to America at eighteen and "worked three years to pay his passage across the Atlantic. His first location was made in Switzerland County, Indiana, and when his labor had paid for his passage, he began working as a farm hand by the month. Three years he spent in that way, and then removed to Hamilton County, Ohio, where he engaged in cultivating a vineyard until 1865. That year witnessed his removal to Erie County, Pennsylvania, where he followed the same line of business until his death." John and his wife had twelve children; nine reached adulthood and five (two women and three men) were involved in some form of agricultural activity, either as a fruit grower or farmer, or as the wife of a horticulturist or fruit grower. John Mottier was well known and active in agricultural activities in the Cincinnati area and published articles on wine-making. See *Cincinnatus; dedicated to Scientific Agriculture, Horticulture, Education and Improvement of Rural Taste,* ed. F. G. Carey (Cincinnati, 1858), the publication of the Cincinnati Horticultural Society, and *Western Farmer and Gardener,* June 1844.

25. The Catawba grape, used for juice, jams, and jellies, is likely a cross of the native *Vitis labrusca* and another *Vitis* species grown mainly on the East Coast of the United States, first discovered in the Carolinas or Maryland. During the early nineteenth century, it was America's most widely planted grape. The Isabella grape, a cultivar also derived from *Vitis labrusca,* is used for table, juice, and sometimes wine. The Cape and Schuylkill grapes mentioned here are likely the Alexander grape, a spontaneous cross of vines discovered in 1740 in Philadelphia where vines had first been planted in 1683. This grape was popularized by William Bartram and widely dis-

Ferguson, B. L. C. Wailes, C. S. Tarpley, and M. W. Philips, Esquires.[26] But, as did a like request made through me by several gentlemen in Kentucky, last season, it met with a decided denial. I much regret that such should be Mr. Longworth's policy. It is not for me to explain it. But I feel confident that a like favor asked by him of any one of the distinguished gentlemen I have named, would have been at once complied with, let the bearer of the request be ever so unworthy.

But I digress. The Isabella has been rejected for some time, as a wine-making grape. Of the other two named, the Catawba ranks first, the Cape second.

The Catawba generally bears a plentiful crop; the grapes ripening well. The juice is abundant, and of good specific gravity, varying of course with the location, soil and season. In very warm and dry seasons, the *must* is richer than during wet ones, and contains more saccharine matter (sugar)— rarely enough, however, to make a good wine without the addition of more, in some form. A rounded bushel of Catawba grapes affords four gallons juice. If this is rich enough to bear up an egg, so as to show any portion of its surface above, it is strong enough.[27] If not, add sugar until it is. From six to ten ounces of New-Orleans brown sugar to the gallon is most commonly

tributed in America by him after the Revolutionary War. From this grape the first commercial wine in America was produced. Affleck fails to mention the scuppernong grape, a variety of the muscadine (*Vitis rotundifolia*) native to the southeastern United States. It was first cultivated in the mid-eighteenth century.

26. William St. John Eliott (1824–58) was a cotton broker, planter, and owner of D'Evereaux and Saragosa plantations in Natchez and Adams County, Mississippi. John Francis Hamtramck Claiborne (1809–84), the "Father of Mississippi History," was a son of General Ferdinand Claiborne and nephew of William C. C. Claiborne, and he lived in Adams County. He was a lawyer, newspaperman, historian, and planter, and a member of the U.S. House of Representatives. William J. Ferguson is mentioned early by Affleck in his article "A Trip Through the South." Benjamin Leonard Covington Wailes (1797–1862) is discussed in a preceding footnote. Colin S. Tarpley and Martin W. Philips of Hinds County were leading stockbreeders in central Mississippi. They were instrumental in activities of the Mississippi State Agricultural Society and often took prizes for their livestock at the Society's annual fair. Reports of the Society and its fair, together with a report of the Agricultural, Horticultural, and Botanical Society of Jefferson College in Washington, Mississippi (by B. L. C. Wailes), appeared in the *Western Farmer and Gardener* in May 1842 (Vol. 3, no. 8: 183–84). All of these men were prominent Mississippi planters (and slave owners) who advocated for agricultural advancement through education, agricultural diversification, improving herds, and scientific methods of farming.

27. Recall that the *must* is the first step in making wine and therefore is a slurry that contains juice, skins, seeds, and stems.

added. It must be put into the vessel placed for the reception of the juice as it flows from the press, and well mixed *previous to the slightest fermentation taking place.* This Mr. Mottier seems to dwell upon very particularly. The grapes are gathered when fully ripe; if there are any unripe and decayed berries on the bunches, they are picked off. They are then placed in a large tub or other vessel and crushed, taking care to bruise none of the stems and seeds that can be helped, as they impart a roughness and astringency to the wine. In Europe the berries are crushed by treading with the feet. Where a clear wine, of the finest flavor and most agreeable aroma is wanted, the *must* is placed at once in the press, previous to its fermenting in the least, and the juice run off through a fine hair sieve into a pipe or hogshead, in which the sugar, if any is found necessary, has been placed; and, after it has been completely dissolved and stirred up, the juice is allowed to undergo a complete fermentation, without being disturbed. When that becomes so far reduced as to render it safe to close up the bung-hole, it is done, and an airhole bored near it. When this is no longer necessary, it is also closed up. The wine is then allowed to stand until it fines,[28] which, unless too rich, it will generally do by the first of March. If not fined then, take the whites of eggs, in the ration of half a dozen to forty gallons, and beat them up effectually; in fact, until they are a mere froth; mix with them some of the wine, and pour the whole into the pipe, mixing effectually—allow the whole then to settle, and, when perfectly fine, rack off into casks or bottles.[29] If the temperature of the cellar be too low, the wine will fine slowly—if too high, it will be apt to sour. About 45 of Fahren. may be considered the medium. The large cellars in Liverpool are carefully kept at 47. If the juice be of great specific gravity—floats an egg very high—it will ferment best in greater quantity—if weak, in lesser quantity, and be less apt to sour.

If you desire to produce a dry, rough but strong wine, without reference to flavor, ferment the *must* before pressing. After the grapes are carefully and equally crushed, leave them in a cask or other vessel, throwing a few folds of blanket over it. Allow the fermentation to proceed until the pumice rises to the surface, begins to crack open, and is covered with white bubbles; the pure juice must then be drawn off from below, and afterwards press the pumice.[30]

If making wine from the Cape grape, the same process is pursued, with these differences: About one pound of sugar may be added to every gallon

28. To fine is to clarify.
29. This process is a fundamental culinary technique for clarifying meat stocks.
30. Pumice is the residue of solids remaining from wine-making process.

of Cape juice. It will most commonly fine itself by New Year's without any artificial means being used. It makes a very agreeable wine.

These hints may be the means of pointing out to you the causes of your want of entire success in your experiments hitherto—if so I shall be gratified. The country would be highly indebted to you for a short account of your attempts at introducing the cultivation of the grape, and the making of wine, in Mississippi; the particulars of your success, and your views of the difficulties you have had to contend with. Your doing so would be the means of extracting much useful information on the subject. If every plantation in the South had a vineyard of sufficient extent to afford each hand a pint or two of ripe grapes per diem, during the season, there is no doubt it would add immensely to their health. The space occupied and the expense of tending them would be a mere trifle. A lack of space compels me to leave your other enquiries unanswered for the present.

I am, my dear sir,

Yours very respectfully,

T. A.

*By the way, during a short trip through part of Ohio in October last, I met with someone, a reader of the *Farmer,* who assured me that he knew the original vine, from which the cuttings here spoken of were taken—that a Dr. Brown sent them, &c.—Other particulars were given me, of which I made a note at the time—but it unfortunately happens that my entire memoranda of that trip have been lost or mislaid. If this should meet the eye of the gentlemen with whom I conversed, I should be glad indeed to hear from him on the subject.

## ON GARDENS AND GARDENING

In the following articles, titled "The Garden" and "The Kitchen and Flower Garden," Affleck addresses specific members of his audience: "Lady Readers," the "Farmer," "Professional Gardeners and Nurserymen." He also takes up the subject of bees. Affleck later expands on the subjects of bees and mulberry cultivation (included here), and he later publishes a book on bees, *Bee-breeding in the West* (1841). Interest in growing mulberry trees was an agricultural fad at the time that did not produce the economic results anticipated. Note that Affleck also mentions support for the "establishment of horticultural societies" (recall the preceding article), for which, with other interests, "we will be found ready if not able advocates."

In his discussion of kitchen gardens, Affleck refers to widely available horticultural books by McMahon and Bridgeman, noting that they give directions by the month. This format, established centuries before in England,[31] was popular throughout nineteenth-century America. Affleck would later edit the *Western Farmer and Gardener's Almanac* (Cincinnati, 1841) and *Norman's Southern Agricultural Almanac* (New Orleans, 1846), which later became *Affleck's Southern Rural Almanac and Plantation and Garden Calendar,* published between 1851 and 1862.

— Ed.

## The Garden[32]

"'Tis a winning thing, sir, a garden! It brings us an object every day;
and that's what I think a man ought to have if he wishes to lead a happy life."
—EUGENE ARAM

The garden is the source of so many enjoyments, both pure and gross, spiritual and sensual, equally gratifying to the tastes of the admirer of nature and her works, and of the lover of the good things of this life, that we are truly surprised it should be so generally neglected as it is. How rare it is to find a *cultivated* garden! If, amid the wilderness of weeds to which we are directed as bearing the name, we discover some little nook devoted to a few favorite flowers, common perhaps, and crowded together, yet free from weeds, and originally arranged with some regard to neatness and taste, we at once recognize it as the work of some fair spirit, whose natural love of flowers thus shows itself. It is a taste natural to woman—not merely from her admiration of the beautiful and chaste, but because in those "loveliest of nature's gems," she finds an object to love and cherish, which repays her care and attention by giving forth an unwearied succession of unobtrusive

31. For an early English example, see Sally O'Halloran and Jan Woudstra (2012), "The Gardener's Calendar: The Garden Books of Arbury, Nuneaton, in Warwickshire (1689–1703)," *Studies in the History of Gardens & Designed Landscapes: An International Quarterly* (2012), DOI:10 .1080/14601176.2012.738472

32. From *Western Farmer and Gardener,* 235–36.

beauties emblematic of herself. But alas! She too often meets but a poor return for all her care; from the want of the necessary knowledge of the plants
she tends, she crowds all in together, the modest and lovely with the showy
and overtopping, the annual and the perennial, until, when she looks for her
abundant harvest of sweets, she finds it indeed but scant.

To give this needful information, to guide this untaught love of flowers, it
is that we propose, when speaking of their cultivation, to address ourselves
more immediately to our

LADY READERS, giving them carefully prepared directions, such as will
enable them to grow, in addition to the common shrubs and flowers, those
beautiful annuals which gardeners denominate as *tender* and *half-hardy,*
with an occasional hint as to the management of green-house plants, roses,
geraniums, &c., which are deservedly such favorites with the ladies, but
which they have generally to mourn the loss of when winter sets in, for
want of a little knowledge of the care they require; and though we occasionally find that *utilitarian mortal,* man, possessed of a leaning *flowerward,* we
much more frequently find him turn from them with perfect indifference to
what pleases him better, a goodly crop of corn or potatoes!

TO THE FARMER, then, we will give articles on the preparation of the
ground for the garden, on the culture of the more bulky vegetables, the
means of saving them over winter, &c., &c., on the raising, or choice, of
young fruit trees, and on the planting and after management of the orchard,
vineyard, &c., &c.

TO PROFESSIONAL GARDENERS AND NURSERYMEN, we do not expect to be
able to furnish much information from our own individual experience, but
we do hope to induce them to make our paper the vehicle of conveying to
each other, and to the world, the valuable results of their own practice and
experience; and in return we will quote all we think of interest to them from
the various British and American gardening periodicals, and more particularly from that valuable work, "Hovey's Magazine of Horticulture," which
we seldom open without finding something new and interesting. Following out the plan so ably commenced in that work, we will publish frequent
"Notes on Gardens and Nurseries," with the view of keeping the *gardening
world* informed as to the state of Horticulture in the west, a matter in which
we regret to perceive, our eastern friends are in complete darkness. We are
promised, on this subject, articles from the pen of one of our few amateurs—
one whose soul is in the task.

TO BEES, as having their home in the garden, we propose devoting a space in each number, in order to induce our readers to pay more attention to the care of those profitable and interesting little busy-bodies, and we would *earnestly* urge it upon those who have had experience and success in their management, to aid us in our efforts by contributing all the information they can on the subject; for even a single fact, told in the plainest possible manner, we will feel obliged, and particularly anything relating to the prevention of the ravages of the bee moth.

For the growing of the mulberry and of silk, the establishment of horticultural societies, and in short everything that we consider useful or interesting to the gardener or the farmer, we will be found ready if not able advocates.

T. A.

———————◆———————

### The Kitchen and Flower Garden—The Shrubbery and the Orchard[33]

This heading embraces a wide field, and one that can by no means be occupied in a single chapter—but the appropriate work for the month, in each department, may be treated of at sufficient length. Our readers must not expect us to inform those of them, who are altogether ignorant on this subject, of all that it is necessary they should know. Let them provide themselves with some such work as McMahon's or better still for those with small gardens—Bridgeman's Gardener's Assistant.[34] We can only remind them of the work needful to be done in each month, with a very short sketch of the best manner of doing it; and keep them instructed in all that is going on, in the way of improvement in the gardening world.

33. Ibid., 335–36.

34. Bernard McMahon or M'Mahon (1775–1816) was born in Ireland, came to America in 1796, and entered into the seed and nursery business in 1802–1803. His *The American Gardener's Calendar: Adapted to the Climates and Seasons of the United States* was first published in 1806 and remained in print for decades, with the eleventh edition appearing in 1857. Considered the first American gardening book, it remains one of the most important because of its content and popularity. Early on, McMahon established a relationship with Thomas Jefferson, and in 1806 Jefferson selected McMahon as one of the two nurserymen to receive plants collected by Lewis and Clark. Thomas Bridgeman (?-1850) published *The American Gardener's Assistant* in 1835. Considered a standard reference, it was widely available and went into multiple editions throughout later decades of the nineteenth century.

If the garden was not well manured and dug over during the fall or winter, let it be done now as soon as possible, but not while the ground is too wet. Leave it rough until needed, in order to expose as much surface as possible to the action of the air and frost. Prepare a few beds, in a sheltered situation for early cabbage lettuce, cress, radish, turnips, peas, celery, tomato, and sow them as soon as possible; some of these may require protection while severe frosts continue.

Dress beds of asparagus, rhubarb, artichoke, sea-kale, &c.

Sow full crops of onions, parsnips, carrots, asparagus, rhubarb, sea-kale, leek, parsley, salsify, spinach, &c. Plant early corn, potatoes, artichokes, rhubarb, cabbage and turnips for early greens. Make permanent asparagus beds. Towards the end of April, plant beans, cucumbers, melons, squash, nasturtium, beets, peas for succession, pumpkins, &c.—sow late kinds of broccoli and cauliflower, cabbage for summer use, celery for full crop, endive, and all the sweet aromatic and medicinal herbs. Separate and transplant all kinds of perennial herb roots, such as mint, sage, &c. Besides the work of sowing and planting the various kinds of seeds above enumerated, all the strongest plants of cabbage, cauliflower, and lettuce must be taken from the hot-beds and frames, and transplanted into the regular beds in the open garden. Attend to such other business in this department as was left undone last month, and see that the garden be help neat and free from weeds.

THE FLOWER GARDEN.—The ground ought to have been prepared last fall; but if not let it have a careful digging now, adding some well-rotted manure.[35] Let it be laid out and planted with shrubbery, as taste may direct. In planting shrubs, be sure to get good plants, true to their kind; have them carefully taken up and as carefully replanted—not more than half an inch, or an inch deeper than they were before. See remarks on planting fruit trees in a former number. Sow some of the more hardy annuals towards the end of the month; and the tender, and half-hardy, in pots or in a hot-bed. About the middle or end of April, plant out amaryllis formosissima, gladiolus psitacinna, tiger flowers, tuberoses, and such other bulbs as may have been pre-

---

35. "Fresh" manure is too strong to apply directly to plants; usually barnyard manure was collected and mixed with vegetative matter and left to "mature" or decompose for a period of time before being applied as a fertilizer or soil additive. Agricultural papers gave a lot of attention to the subject of manure, its characteristics, value, chemical content, and proper application.

served dry through the winter.[36] See article on the Dahlia, and on the cultivation of annuals. There are some annuals, which, though they should have been sown in the fall to come to full perfection, will still do well sown now.

THE ORCHARD.—Where those trees that require it, have not yet been pruned and cleaned, let it be done now. This is the proper season for planting all kinds of fruit trees, vines and shrubs. See article on the vine. Prune and plant currant bushes, gooseberry and raspberry—the two former, if along the border, six feet apart; and if in a separate compartment, which is the better way, at about the same distance, that they may be tended between. The raspberry, if in a separate compartment, may be planted in rows six feet apart, and the plants two feet apart in the rows; to produce an abundance of fine fruit, they ought to be in a piece of good ground, and be well tended with the plough and hoe. There is no plant more tender or difficult to transplant than this. If taken up in a cold, dry day, and their roots exposed one hour, there is little hope of their doing well. Let them be carefully taken up, in a moist, dark day, and be immediately replanted. The Ohio ever-bearing is the finest variety; and next to it, the red Antwerp and the native purple. Grafting may now be performed with safety and success. Cuttings and suckers should be planted; as also fruit stones and kernels. See article on the strawberry.

T. A.

36. *Amaryllis formosissima,* Aztec lily, more commonly known as *Sprekelia formosissima,* is native to Mexico and Guatemala. *Gladiolus psittacina* is parrot gladiolus; tiger flower is *Tigridia,* shell flower, also native to Mexico.

# ON BEES

Bees have been the subject of fascination, veneration, and study since ancient times. Throughout the world, bees acquired symbolic associations with industry, community, organization, discipline, and intelligence, and their product, honey, became known as the "food of the gods." Since most agricultural products reproduce through pollination (and therefore through the action of bees), connections among agricultural success, industriousness, hard work, and knowledge of bees are evident.

Beehives were important symbols for eighteenth-century secret societies, such as the Masonic orders, and in nineteenth-century America they appear in the iconography of several religions (notably the nineteenth-century Mormons), governments (most seals of American states have agricultural images, and both Arkansas and Utah feature beehives), and agricultural newspapers such as *The Cultivator* and the *American Agriculturist*.

Affleck's interests and writings cover a broad range of topics from animal husbandry to agriculture, so it is no surprise that a topic about which he wrote extensively was bees and beekeeping. Four articles appear in the *Western Farmer and Gardener* in 1840 and 1841, followed by publication of his "little work devoted to this subject," *Bee-breeding in the West* (Cincinnati, 1841), as he notes in the fourth installment of his articles in the *Western Farmer*.[37] This pamphlet-sized publication (fewer than seventy pages) elaborated on the topics he discussed in his *Western Farmer* articles.

Judging from the sources Affleck quotes in his articles, he was well-versed in the writings of European and American apiarists, and he obviously had personal knowledge of beekeeping. Combining the two enabled him to write convincingly about his subject and to offer practical suggestions for inexperienced readers. In the following articles, note Affleck's connection in the opening paragraph of his first installment linking success in keeping bees (and by inference, any agricultural endeavor) and the suggestions found by subscribing to a "Farming Paper" such as the *Western Farmer:* "your successful bee breeder is always a subscriber to some one or more of the '*Farmer's friends.*'" Noteworthy too is Affleck's economic argument for keeping bees, found in the second installment. An image is given in the third article when discussing how to house bees, and the last article deals with how to prepare for swarming bees. In the closing paragraph of the last article, Affleck notes that his aim for

---

37. Limited edition reprint (2011); see http://openlibrary.org/works/OL16458781W/ Bee-breeding_in_the_West.

publishing *Bee-breeding* has been to "issue a useful and cheap work, adapted to the necessities of our farmers—avoiding technicalities, and bringing the subject within the comprehension of all." This characterization might well apply to the entire body of his published works.

—Ed.

## The Bee [38]

To the great and yearly increasing difficulty experienced by bee breeders in their efforts to save their bees from the destructive *moth,* we must ascribe the scarcity of *honey* in the west. Some there are, who, by neatness and care, and by studying the nature of both the bee and their enemy, and by adopting such as they find answer, of the various suggestions they meet with in their Farming Paper (and your successful bee breeder is always a subscriber to some one or more of the *"Farmer's friends,"*) have succeeded in accumulating a stock of those useful and profitable insects. Having long studied the most efficient means of increasing and saving our bees—of keeping off the moth, and of removing a share of the honey without disturbing its legitimate owners; and having succeeded beyond our expectations in these necessary objects, we are anxious that our reader should be put in possession of our methods and its results. But in order that we may be understood, we will first give a short familiar sketch of the natural history and economy of this remarkable tribe of insects—the *"apis mellifica"* of naturalists—and is so doing, divest our language, as much as in our power, of technicalities.

It will be necessary, to enable the reader to understand the functions or duties of each division or kind of bee contained in the hive, to enter somewhat minutely into their separate and comparative physiology, which will comprise their singular division into sexes or "modifications of sexes"; their food, secretions, mode of breathing; their external senses, and their instincts. We will try to clear away those numerous errors which successive authors and observers have disseminated, and by comparing the mass of curious and interesting facts accumulated by Huber, Reaumer, Swammerdam, Wildham, Hunter, &c. &c., and by various writers in our own country,

38. From *Western Farmer and Gardener,* 241–43.

Dr. Thatcher, Weeks, Kelsey, &c., do our utmost to arrive at the truth.[39] We will then follow them in their different labors, from the time the young swarm has settled in its new house; the structure of their hives, their singular and systematic architecture, the rearing of their progeny, and the issuing forth of new swarms; the massacre of the drones, when no longer necessary for the impregnation of young queens; the fecundation of the eggs, which, until cleared up by the observations of Huber, was involved in the deepest obscurity. After having thus possessed the reader with some general acquaintance with the history of these insects, we will proceed to speak of the different and most common mode of managing them, explain wherein we think them erroneous, and give a sketch of such plans as have come under our observation, which proved successful preventives to the ravages of their enemy, the *bee moth,* and which seemed to us best adapted for insuring their health and increase, and enabling their owner to remove the honey without disturbing or injuring the bees; and lastly, we will give a full description and explanation of the method we hinted at, by which we ourselves have almost entirely succeeded in attaining those most necessary and desirable objects.

In our sketch of their natural history, we will of course give precedence to the queen, as mother of the hive. To begin with the egg—that from which she is hatched, is precisely similar to those which produce the working bees; the larva or worm comes from it in the same manner, and does not differ from that of the worker. But the cells in which the eggs intended to produce queens are deposited are large, being above one inch deep, one-third of an inch wide, and their walls, which are formed of wax hardened by a mixture of propolis, are nearly an eighth of an inch thick; they are most commonly built on the edge of some of the shorter combs, but occasionally in the very centre of the hive. They vary in number from three to four, to twelve or fourteen. Their form is an oblong, resembling that of a pear; their position is always vertical, so that when they arise from amidst other cells, they are placed against the mouths of those cells and project beyond the common surface of the comb. They are perfectly smooth on the inner surface, while their outer side is covered with a kind of hexagonal fretwork, as if they were intended for the foundation of regular cells. The eggs being deposited in these cells are hatched without requiring any particular attention from the

39. Those mentioned here are eighteenth- and nineteenth-century scientists from Europe and America who had studied and written about insects in general and bees in particular. Their works were the authoritative source books of the period.

bees, except that of keeping up a proper temperature, in which case the larvæ appear in three days, and have the appearance of small white worms without feet, coiled up at the bottom of the cells. From this time the attention of the nursing bees is much more incessantly given to the royal larvæ than to that of the workers or drones, and they are fed with what appears to be a much more stimulating food than that given to the others; it has not the same sickening taste, but is somewhat acid, and is given in such quantity, that a part always remains in the cell after the queen leaves it. It is thus *"forced,"* as it were, into a more full and rapid development of all its organs, and in five days from its being hatched, it is ready to open its web preparatory to its transformation, and the bees enclose it by building up a wall at the mouth of its cell. The web is completed in twenty-four hours, and after remaining in a state of inaction for two days and a half, it becomes a pupa. It remains between four and five days in this state; and thus, on the sixteenth day after the egg has been laid, it has produced the perfect insect. When this change is about to take place, the bees gnaw away a part of the wax covering of the cell, till it at last becomes pellucid [transparent] from its extreme thinness. The queen, although perfectly formed, is not always allowed to leave her prison; but of this we will speak again, and at present suppose that the old queen having gone off with a swarm, the presence of a young one is required, and she is accordingly liberated by the workers and comes forth to perform the duties of her station. She is larger than any of the other bees, with mandibles or teeth of smaller size than those of the workers, though larger than the drones; wings much shorter, extending little, if any, beyond the third ring; proboscis shorter, and her stinger short and curved.

The drone, which is admitted to be the male of the species, is characterized by a thicker, flatter body than the worker, rounder head, a more bluntly terminating abdomen, within which is contained the mail [male] organs of generation, and which take the place of a stinger, this weapon being denied them. The eggs from which the drones are hatched are deposited in cells somewhat larger than those appropriated to the production of workers, and are in proportion to those of the latter, as one to thirty. They require about twenty-four days from the laying of the egg til their becoming a perfect insect.

The workers, or as they are most commonly, though somewhat incorrectly called, *neuters,* comprise the third class; are smaller in size than either of those we have just described, proboscis or trunk more lengthened; the structure of their legs and thighs peculiar, having a concave or hollow space on the middle joint of their hinder legs, surrounded by a row of hairs. In this "basket," as it has been termed, they carry home the pollen of flowers, usually called *bee-bread,* which they first collect by rolling themselves on it in the blossoms, and then brush it off with a small brush or pencil of hairs which grows on the *tarsi* or last joint of the leg, knead it into a ball and place it in the "basket," where the surrounding hairs retain it in its place. Till within a very few years, the working bees have been considered as *neuters,* or *mules*—animals deprived of sex. It is now proved beyond a doubt by the observations and experiments made in various parts of Europe, confirmed by those of Mr. Huber, of Geneva, that they are in reality *females,* having all the necessary though *undeveloped* organs, which, we have seen, in those larvæ intended for queens, were *forced* into a perfect or mature state. Of the fact of workers having become impregnated, and, where the hive happened to be destitute of a queen, having laid eggs, which, however, invariably produce *drones,* we will speak at some future time, this article being already sufficiently lengthy.

T. A.

———◆———

## The Bee—№ 2 [40]

No branch of rural economy yields so great a return of actual profit and of rational amusement, as the cultivation of bees. How surprising, then, that so little is done towards its improvement! As I before remarked, to the ravages of the bee-moth, we are to look as the principal cause. Though we may find another equally great, in the unwillingness so many of our farmers manifest, to step out of the track they have so long followed. They say, bees now require too much trouble; if they could be kept as easily and do as well in their *old Sycamore gum,* or big box, *made after the young swarm* has left the parent hive and settled, as they were wont in the early settling of the country, there would be some satisfaction in keep them—but as for all this pains-taking in the manufacture of boxes; this continued cleaning of stands.[41] Yet these very men who look on bee-tending, the planting and cultivating of orchards, the care of poultry, or the breeding and care of silkworms, as hard and troublesome work, will toil all summer, without a com-

---

40. From *Western Farmer and Gardener,* 255–56.

41. "Sycamore gum" refers to a hollow section of tree used to capture swarming bees or to house a beehive. Bees swarm in the spring or fall when the colony gets too large. The existing queen leaves in search of a new home, taking part of the hive with her to a new location (often

plaint, raising a crop of corn, which will yield them a comparatively trifling remuneration. They acknowledge stock-raising to be the most profitable branch of agriculture, and envy those who have the means of going into it—yet here is a stock for which they have "unlimited right of pasturage"—that they may turn out to range at will, without danger of their being taken up as "strays," or complained of as "breachy"—and which require no exorbitant outlay of capital to commence the business, and certainly sufficiently neglected to leave room for competition! There is more to be feared from embarking in the raising of "Durhams" or "Berkshires!"[42]

As to profit to be derived from bees, much has been published—let me give my own experience.

About Christmas, 1836, I purchased two stands at a cost of three dollars each. This was in western Indiana—they would have commanded that sum here. They were anything but good, being in the common old rough boxes, full of cracks, with several entrances large enough for a mouse. All this I remedied as well as I could, and next winter my stock had increased to four. I was much from home during the season of swarming, so that three fine swarms went off during my absence, there being no one at hand to save them. These four I kept undisturbed; and the next fall, when I left the country and gave away my bees, the account stood thus—though I may again remark, that during absences from home in swarming time, I again lost five swarms, which I regretted much, but could not prevent.

---

a hollow in a tree or a limb) and leaving behind queen eggs, one of which will become the old hive's new queen. This swarm, a large, teeming ball of bees, will form in a hollow of a tree trunk or completely cover a tree limb. It was common in nineteenth-century rural America to collect these swarms and house bees in sections of the naturally hollowed tree trunks ("gums") of sycamores (*Platanus occidentalis*), sweet gums (*Liquidambar styraciflua*), or black gums (*Nyssa sylvatica*). These common native hardwoods often have hollow trunks and limbs used as nesting places for animals and insects. Their flowers and sap, particularly the sweet gum's, are attractive to bees. Manmade gums, usually about twelve to eighteen inches in diameter and sawed into 24- to 30-inch-long sections, were made from trees with existing beehives or created in a hollowed tree trunk enhanced by cleaning and smoothing the interior so that sap exudes, attractive to the bees. Swarms could be captured by shaking the swarm off the limb or by scooping the bees by hand into the gum, thereby creating a new hive. Gums could house the bees or be re-used if the bees were subsequently relocated to wooden boxes, which were often placed on top of gums. This process of moving and creating new hives, still practiced in Appalachia, is likely the same method employed in the nineteenth century. See *Foxfire 2,* ed. Eliot Wigginton (New York: Anchor Books, 1973), 32–47.

42. Durhams are cattle; Berkshires, swine.

I had, then, in the fall of 1838, ten fine swarms, nine of them arranged as I shall hereafter describe, and three of the young stands, of that same season, having sent off each a swarm. Not wishing to disturb my young stands, and the old boxes being an eyesore to me, I drove the bees from one of them the second summer and took the honey. In doing so, I have no doubt I accidentally killed the queen, as the bees, that evening, left the box I had put them in, and quietly distributed themselves among the other hives—so that I still had the bees, though my number of stands was reduced. Thus, from an outlay of $6, I had, in two years, and at an expense of time and trouble amply repaid by the amusement and instruction they afforded me, ten stands, worth, from the order they were in, at least, even there, $4.50,

$45.00

And from the one I robbed, I got 35 lbs of honey, say

7.00

Total $52.00

What *crop* will yield such a profit? And then, had I been able to be about, during the swarming season, so as to save all that came off, I have no doubt my stock would have been nearly doubled.

The singular instinct and intelligence of this insect, was displayed very clearly in the conduct of the swarm I had driven from their old hive to one of my pyramidal boxes. (Of the method and proper season of *driving,* I will speak again.) It was after sundown at least an hour, when I had got them transferred to their new box, and set on one end of the same plank on which were five others. I saw, in a short time, that something unusual was the matter, and stood by to watch the result. For half an hour they remained in the hive, and at once, as if by a preconcerted [*sic*] arrangement, they marched out in a continued stream along the front of the other hives, into each of which a smaller stream branched off from the main one, gradually lessening it until the *van* marched into the last hive! In two or three minutes the box was empty—not a bee left, and all was quiet. So far as I could perceive, about an equal number entered each hive. Next day everything went on as usual, except that they seemed more irritable than common.

The subject of the generation of bees has long been, and still continues to be with many, a disputed point. To me it seems that the experiments of Huber and their results set the matter forever at rest. It will be unnecessary for me here to give all the different theories at length; suffice it that some assert that a sexual union takes place between the queen and the drone, within the hive, though they could only state that result of their observation to be an indistinct and transient junction—others could see nothing of this, but insist that the queen is a hermaphrodite, having within herself the powers of both sexes; and proving, moreover, that on her being confined alone with two drones, she turned on them, on their approaching her, and killed them on the spot. Swammerdam—and many believe with him—contends that the impregnation takes place from a certain *aura,* proceeding from the bodies of the males, which needs must be numerous, that it may have sufficient power. Many other such doctrines are advanced, the most plausible of all, being that of the eggs being first deposited in the cells by the queen and there impregnated by the drones ejecting the seminal fluid over it, as we know to be the case in the spawn of frogs and of fishes. Some have even insisted that they have seen the drone in this act! But this is disputed by those whose close and continued observation enabled them to state, that it never does take place. And it is also disproved by the fact that eggs are deposited by the queen *after* the destruction of the drone, and are hatched *before* the existence of a single drone in the spring.

It seems to have been reserved for Huber to determine this mooted point. In one of his experiments he removed all the females from a number of the hives, giving to each a queen taken the moment she came to maturity. He then removed all the drones from one division of these hives, letting them remain in the others. He then adapted to each hive a glass tube for an entrance, so small that no drone could pass through it, but large enough for the common bees—thus confining the queen alone with the neuters and her seraglio of males, in the one division; and with neuters alone in the other. To his surprise, all the queens remained sterile! He continued his experiments, diversifying them in every possible manner, and proving at last that the queen bee was impregnated by an actual union of the sexes, as in most insects and in the larger animals; but that this never took place within the hive.

T. A.

———————◆———————

### The Bee—№ 3[43]

Being fearful of fatiguing the general reader, by treating of the more dry parts of my subject, at too great length, I shall aim at giving as great a vari-

43. From *Western Farmer and Gardener,* 270–72.

ety as possible in each number—this must be my excuse to those, who think that I ramble too freely from one division of my text to another.

Having several urgent requests, from individuals who are anxious to improve their mode of managing bees, to give them my method now, that they may prepare their stands and boxes during the winter, I shall proceed to do so; and, that I may be the more easily understood, have had a cut [an illustration or image] prepared illustrative of the subject. I may premise that I lay no claims to originating the subtended hive, but only to improving on the suggestion of M. Ducouedic, a celebrated French apiarist; and adding to his plan a stand or shelf of my own; which, with a reasonable share of attention, will secure the bee from the moth, and from hornets, and other such enemies; and enable the proprietor to remove all the honey each hive can spare, without inconvenience or annoyance to the inhabitants.[44] The great and insuperable objection to bee-palaces, rooms fitted up in the garret, large boxes with small one or glasses fixed on top, and the various other like plans, too generally adopted, and always ending in disappointment, is, that they all aim at compelling the bee to work contrary to its natural habits, instead of guiding and assisting those habits. One object aimed at, is to prevent their swarming—an ill-judged attempt and poor economy, and I shall prove before I have done. Another is, the convenience of getting honey at any time—to attain this, boxes or glasses are placed on *top* of the main hive, with the idle hope that the bee will reverse its usual course, and work *up* instead of *down*. They may do so once, and yield their proprietor some very fine honey, in nice, white comb; but rarely work up a second time, only using the empty box as a withdrawing room, in very warm weather! Such additions, too, form capital breeding places for the moth.

But of the hive itself first—of its advantages over all others I am acquainted with, I will speak again. In the accompanying cut, we have the stand and the subtended hive in its different stages, Nos. 1, 2, and 3. The stand is made of a piece of two inch pine plank, what carpenters call, "*clear stuff*" [i.e., free of knots]; length 24 and breadth 18 inches; 8 inches from one end, and 2 from the other and from each side, is marked a square of 14

44. Pierre Louis du Condeic de Villeneuve's *La Ruche Pyramidale: Methode simple et Naturalle* (Paris, 1813; American edition: Philadelphia, 1829) discusses his methods for building a beehive, and these methods are discussed at length in Robert Huish's *A Treatise on the Nature, Economy, and Practical Management of Bees* (London, 1817).

inches; from the outside of this square, the board is dressed off, with an even slope, until its thickness at the front edge is reduced to half an inch, and at the other three edges to about an inch. The square is then reduced to 12 inches, in the centre of which is bored an inch auger hole;—to this hole the inner square is also gradually sloped to the depth of an inch—thus securing the bees from any possibility of wet lodging about their hive. There will then be a level, smooth strip of one inch in width, surrounding the square of 12 inches, on which to set the box or hive. Of course, it is not absolutely necessary to use these precise dimensions—I am only giving those that I have found answer best. Four inches from the front edge of the stand, commence cutting a channel, an inch and a half in width, and of such a depth as to carry it out, on an even slope, half way between the inner edge of the hive and the ventilating hole in the centre. Over this, glue a strip of wood, after fitting it neatly, dressing down even with the slope of the stand, so as to leave a tunnel an inch and a half in width, by a quarter of an inch deep. Over the centre hole, nail a scrap of sieve wire, to prevent the entrance of insects; and over the outlet of the tunnel, hang a small wire grate, in such a way as that it can either be thrown back to permit the exit of the bees, or fastened down to keep them at home in clear days in winter. For feet to the stand, I let in four of what are called double ten penny nails, or 20d spikes. The lower end ought also to be planed smooth, and the whole should have two coats of white lead paint, sometime before it is wanted, that the offensive smell of the oil may be dissipated before the bees are brought in contact with it.

The boxes are formed of good pine or poplar plank, one inch thick, and twelve inches (or a cubic foot) in the clear, every way. They must be planed inside and out, except the top, which will be left rough on the lower side, that the bees may be able to attach their comb to it. They should be neatly put together, the edges joined, so as to fit perfectly close, and leave no crevices for the moth to insert its eggs. The more effectually to prevent this, I find it a good practice, before nailing the sides together, to put a little thick lead paint along the edge, which renders it completely tight. The top ought not to be served so, but put on as closely as possible with six or eight screw nails, that it may be removed to facilitate the taking out of the comb. In the top is bored four, inch and quarter auger holes, to allow of a communication between the boxes, in the hives № 2 and 3; but in № 1 these are plugged up quite tight. I should also give the outside of the boxes two coats of white lead paint, puttying up every crack or crevice before laying on the second coat—it is in such places that the moth lays her eggs, and when a box is carelessly made and full of cracks, or stands so long on the shelf as to become so, it is impossible to prevent the destruction of the bees contained in it.

Many, I have no doubt, will shrug their shoulders at the idea of taking all this pains with the bee-hive—to the care of such, I would not recommend the apiary. Any man who can handle a plane, a chisel and a hammer, can prepare both the shelf and box, as I have described, without applying to the carpenter, and thus profitably occupy many a wet, dreary day, that otherwise would hang heavy on their hands, and afford them a pleasant relaxation, after even reading had become irksome. Then, think of the profit!

When a young swarm comes off, in the spring, take one of these boxes, after neatly plugging up the holes in the top, put the swarm in it (of swarming, &c., anon,) and in the evening set it on the shelf in the apiary or bee shed. If the swarm is at all a strong one, they will come very near filling one box—I have had swarms, from the subtended hive, that came near to filling two. It rarely occurs, however, that a single swarm is too numerous for one. If they are not so crowded as to require the immediate addition of another, let them remain undisturbed in the first one (№ 1) for a week or ten days; when, by gently raising the box some evening, it will be found that they have nearly if not entirely filled it with comb; in which case a second must be added, by an assistant lifting the one in which the bees are, and another, after carefully sweeping off the stand with a little broom, slips the other one under, the holes in the top of which have been left open. The bees next day continue to work down through the second box, as quietly as if nothing had been changed, and in two or three days, if the season be at all favorable, they will make a considerable quantity of comb in it. If the season happens to be a poor one for the bees, and the swarm originally rather weak, the hive № 1 may be found sufficiently large for them the first season, and they must be allowed to winter in it, not forming a № 2 of it before the following spring. This is always to be regretted, and should be guarded against, by putting two such swarms together, so soon as it is seen to be requisite.

T. A.

## The Bee—№ 4 [45]

In my last article on the Bee, at page 63, I described the subtended hive, recommended its adoption, and partly explained its use.

To those who have bees in the old, clumsy boxes or gums in common use, I would recommend the economy of allowing them to remain as they are, and not attempt to drive, but save the young swarms, as they come off, in boxes on the subtended plan. The old gums ought to be carefully examined and repaired as well as circumstances will permit, turning them up and removing all appearance of the moth, and cutting away the old, black empty comb. The bees will thus have some space afforded them, where they require and want it, and will throw off stronger swarms in consequence. Early in May, if the season be favorable, they will begin to throw off their first swarms—varying as to time in different locations. In the meantime preparations ought to be made for their reception in the apiary; the hives and stands should be got ready, and a shelter erected; an open shed answering very well, so arranged as that they shall not be exposed to the full blaze of the hot summer's sun. A few trees in front of a moderate growth, and trimmed up sufficiently to prevent the bee being entangled in the branches in returning loaded to the hive, will be found of great assistance in swarming, as the bees will almost invariably settle on them, and remain from fifteen minutes to an hour, affording abundance of time for saving them, if the proprietor has everything ready. But if they are allowed to go off a second time, there is little or no hope of saving them. All the noise usually made by beating tinpans, ringing bells, etc. etc., is perfectly useless; though it may sometimes happen on a very clean, warm day, that they will show an unwillingness to settle, in which case they may be induced to do so, by casting a few ladles full of water amongst them, and even by firing a gun near—the concussion throwing them into confusion, and inducing them to settle. When they are all quiet, take a box, the holes in the top of which have been carefully plugged up, and after seeing that it is clean and set, shake or sweep the young swarm gently into it. The face and hands ought to be protected by a veil and gloves; for though bees are not inclined to sting when swarming, if gently handled, yet a chance-sting inflicted on a tender place, will discompose the most firm, and probably occasion twenty more—for it is well known, that the odor of

45. In its original printing. The article as reprinted here is taken from the collected edition (1850) of *Western Farmer and Gardener,* 331–32.

the poison is very strong, and immediately perceptible to the bees, having a most irritating effect upon them. If the outside of the boxes have not been painted sufficiently long to allow the smell of the oil to dissipate, it will be well to rub it over, inside and out, with some sweet herb, such as balm, or even with hickory leaves, which will make them better satisfied with their quarters. So soon as they seem to be somewhat settled, place them on a table or stand of some kind, to allow the stragglers to collect; and in the evening set them on their stand in the apiary.

White clover and mignonette have always seemed to me to pay a handsome profit, as affording food for bees, on the space they occupy and the trouble incurred.

About the 15th of April, I will issue a little work devoted to this subject, entitled "Bee-breeding in the West," containing much more extended particulars of the management of this insect, on the plan of the subtended hive; as also their protection from the moth, and the other enemies they have to contend with, *here in the West.* My aim has been to issue a useful and cheap work, adapted to the necessities of our farmers—avoiding technicalities, and bringing the subject within the comprehension of all.

T. A.

# ON THE CULTIVATION OF MULBERRIES

From the eighteenth century onward, there was interest among entrepreneurial Americans in developing a silk industry using mulberry trees for silkworms, since at least two native mulberries, *M. microphylla* and *M. rubra,* are widely found throughout North America. When efforts with the native mulberries proved ineffective, the Chinese mulberry, *M. alba,* was introduced, but it proved disappointing as well. Nevertheless, interest remained in developing this industry, and periodically, new efforts were initiated by speculators, fanned by claims of ease of cultivation and quick profits on investments. Agricultural newspapers were ideal vehicles to encourage these speculative efforts, and one such episode happened in the 1830s and early 1840s, about the time Affleck was writing the following piece.

Alice Morse Early, writing about the Linnæan Botanic Garden and Nurseries in Flushing, Long Island, in the early twentieth century, records how the country's most prominent nursery was involved. This nursery, started by Robert Prince in 1730, survived the Revolutionary War and eventually passed to William Prince. It was

a centre of botanic and horticultural interest for the entire country; every tree, shrub, vine, and plant known to England and America was eagerly sought for; here the important botanical treasures of Lewis and Clark found a home. William Prince wrote several notable horticultural treatises; and even his trade catalogues were prized. He established the first steamboats between Flushing and New York, built roads and bridges on Long Island, and was a public-spirited, generous citizen as well as a man of science. His son, William Robert Prince, who died in 1869, was the last to keep up the nurseries, which he did as a scientific rather than a commercial establishment. He botanized the entire length of the Atlantic States . . . and sought for collections of trees and wild flowers in California with the same eagerness that others there sought gold. He was a devoted promoter of the native silk industry, having vast plantations of Mulberries in many cities; for one at Norfolk, Virginia, he was offered $100,000. It is a curious fact that the interest in Mulberry culture and the practice of its cultivation was so universal in his neighborhood (about the year 1830), that cuttings of the Chinese Mulberry (*Morus multicaulis*) were used as currency in all the stores in the vicinity of Flushing, at the rate of 12½ cents each.[46]

By 1839, speculation in mulberry plants was rampant throughout the United States, and demand for them far exceeded supply, with entrepreneurs buying and selling trees they did not yet have. Silk factories were established in the southern states, and at least two journals appeared advocating the industry, the *Southern Silk Journal and Farmers' Register* (Columbus, Georgia, 1839) and the *Southern Silk Manual and Farmer's Magazine* (Baltimore, 1838–39). The bubble burst in the fall of 1839, and many lost money. Judging from dated articles that appear before and after, Affleck's piece must have been printed in late 1840. Affleck says nothing of the economic speculation that had consumed many but instead sets out to dispel notions that the cultivation of mulberry trees for silk production was somehow "effeminate" and therefore unworthy of "the attention of *men!*" Clearly he remained optimistic about

46. Alice Morse Early, *Old Time Gardens Newly Set Forth* (New York: Macmillan Company, 1902), 26–27, http://www.gutenberg.org/files/39049/39049-h/39049-h.htm#Page_27, accessed June 10, 2012.

the mulberry's potential for economic development, and by gathering a "collection of facts" from throughout the country, he hoped to reassure his journal's subscribers of the economic viability of growing mulberries for silk production. In spite of economic uncertainties, interest in silk culture continued in the early 1840s in the American South, where it was seen as an industry that could employ aging slaves and child labor. Nevertheless, the silk industry in America never was profitable.[47]

—Ed.

### Save and Cultivate Your Mulberries: Success in Silk Growing[48]

We have made a collection of *facts*, from various papers, as to the success in silk growing in different parts of the country, which we think sufficient to warrant our saying, that there is no doubt whatever of its becoming of equal importance with cotton growing, and far more than sufficient to counteract the evil tendency of such articles as one that appeared in a recent number of one of our best agricultural papers, the "Farmer's Cabinet," where the principal argument made use of was, the *effeminate character* of the employment rendering it unworthy of the attention of *men!*

To those who have large quantities of the *Morus multicaulis,* we would say, do not allow them to be destroyed or wasted; there is no doubt of their being in moderate demand and at fair prices; determine on some price, say eight or twelve cents each for good trees, and keep them rather than part with them for less; by no means *give them away*—your doing so will benefit no one, as those who get them for nothing, most assuredly will not value them. Let those who have trees and ground to spare, set out a permanent plantation of an acre or two this season; our word for it, such a plantation will ultimately pay well.

Mr. Gill's letters speak volumes, and ought to be carefully read by all.

In the neighborhood of Nashville, Tennessee, much has been done; the spirited efforts of a few individuals there, if they go on as they have begun, will render Nashville one of the headquarters of the silk business in the West.

Various individuals in this vicinity have done a little, and almost all have succeeded so well as to be making arrangements to go into it largely this next season.

47. Lewis C. Gray, *History of Agriculture in the Southern United States to 1860* (New York: Peter Smith, 1941), 2:828–29.

48. From *Western Farmer and Gardener,* 293–95.

In Kentucky and Indiana much more has been done than is generally known or believed.

The Burlington Silkworm record says:—

"Mr. Daniel Spaulding, of Hancock, N. H., is now preparing to put up our improved Frame in his cocoonery.

Mr. R. Shore, of Morganville, Va., has built a cocoonery large enough to feed 300,000 worms.

Mr. John B. Hart, of Scott's Ferry, Va., writes us for information as to cost of delivery frames at Richmond, stating that he has some expectation of building a cocoonery this winter, 54 by 27 feet.

Mr. L. A. Spaulding writes from Lockport, N. Y., as follows:—

'I have paid some attention to the silk business, and had foliage sufficient to feed a large number of worms the past summer, but could not get the eggs. I have not a supply. In feeding, I have had good success; and in reeling, beyond my expectation. The whole operation is so simple that its success is no longer doubtful. It is settled that this country will supply itself with a large portion of its consumption in a few years.'

From the Rev. John Foster, of Lebanon, Ohio, we have received the following:—

'I have been trying to make a little silk now for two or three years. I have done as well as I expected, according to my means. Indeed I have not attempted any thing but to learn the habits of the worms, until the past summer. This summer I fed all I could get, which was very few. They did well, except an early brood which was fed upon but native mulberry.'

From Louisiana we have accounts of several gentlemen embarking in the business. Great success may be expected in that delightful climate.

Mr. Frederick Brownell, So. Westport, Mass., obtained 15 bushels of cocoons the present season, from about 60,000 worms.

Mr. N. E. Chaffee, of Ellington, Conn., writes as follows:—

'I have fed worms this season; obtained 20 bushels of cocoons, though fed wholly on board shelves. By what I can gather from the description of your frame in the Record, I think very favorably of it.'

The following extract of a letter from Mr. John Iredell, of Enoch, Monroe Co., Ohio, is strikingly characteristic of his praiseworthy perseverance:—

'We have been engaged for three years past in feeding a few silk worms, from 10 to 30,000, on the wild mulberry, and have produced some excellent silk.'

At Patriot, Ia., a silk growing company has been established, but is not yet in operation. From Mr. J. B. Taylor, of that place, we learn the following particulars, obtained by him from Mr. H. Huxley, of the success of the latter in rearing worms during the past summer:—

'The building occupied as a cocoonery was an unfinished house, floors not laid, and one side without weather-boarding; consequently, the worms were exposed to all the changes of weather. The shelves were made of rough one-inch boards, and the worms placed upon hurdles made of cotton cord. The alpine seedling was used exclusively for feeding, the multicaulis having been layered in the spring, did not produce leaves in season for the first crop of worms, and in the latter part of the season eggs would not be obtained. The cocoons raised will make about twenty-five pounds of sewing silk, at a low estimate. A small reel and spinning mill were constructed, and a very fair and saleable article of sewing silk was produced, which is pronounced by our merchants to be equal to any they can purchase.'

Mr. Taylor himself adds,

'I have a specimen of the silk now in my possession, and can bear testimony to the excellence of its quality. In this section of country it is worth $12 per pound. Mr. H. reeled, spun, and dyed it himself; the spinning mill is of his own invention and construction.'

This is a remarkable statement altogether. No doubt Mr. Huxley is entitled to the highest praise every way; but, so far as related to his cocoons, we believe a great portion of his success was owing solely to *luck*. The best evidence of his mechanical and manufacturing skill, appears in his having produced sewing silk with his own hands, on a machine of his own making, that sold for twelve dollars per pound. Truly, we may expect great results from the enterprise of such a man as this.

And from the New England Farmer:

'One person, this year, has been experimenting upon one-quarter of an acre of mulberries, with the view of testing Mr. McLean's experiment of last year, and also to convince his incredulous neighbors. He will probably have nine to twelve and a half pounds of raw silk from that one-quarter of an acre, worth five dollars per pound, say fifty dollars; whereas, if the same land had been planted with corn, he might have had seven bushels, at seventy-five cents—five dollars twenty-five cents; or, if with wheat, might have had enough to make one barrel of flour, worth six dollars twenty-five cents. It costs no more to transport a pound of silk to market, worth five dollars, than a pound of flour, worth four cents; and while it requires at least *six months* to mature a crop of wheat for the market, *six weeks* are sufficient for a crop of silk; and after deducting every possible expense between the culture of a grain and a silk crop, is not the difference of profit so far superior as to encourage some passing notice of silk, and especially when an inferior soil is adapted to the growth of the mulberry?'

We add, from the Urbana Western Citizen,[49] the following additional particulars:—

'We were not aware, until very recently, that any of our citizens had turned their attention to this subject. Our readers have already been advised of the fact, that Mr. *Lapham,* of Salem township, received a premium of $5 from the country treasure, being the bounty offered by the Legislature for every 50 lbs. of cocoons raised in the State. A few days ago Mr. *Kidder,* of Rush township, presented us with a skein of sewing silk, of his own manufacture, which, for beauty and durability, will favorably compare with the eastern or foreign make. Mr. Kidder informed us that he raised 80 lbs of cocoons during the past season—the bounty on which was $8. This is all manufactured into sewing silk, and sold at a profit. The proceeds of the past season will amount to about $200; and the principal part of the labor is performed by a young lady in the family. This is a good beginning.'

These are encouraging statements, and such as we hope will confirm the waverer, convince the doubter, and silence the sneerer.

T. A.

# OBSERVATIONS ON LIVESTOCK IN KENTUCKY

Two lengthy articles by Affleck appeared in early 1841 in which he reports on the several men he met and farms he visited in Kentucky and the general condition of livestock such as cattle, pigs, sheep, and horses. By this time, he was junior editor of the *Western Farmer,* and this report on livestock conditions in Kentucky is the first of several on general agricultural issues in other regions in which he traveled. Much of this two-part article is devoted to discussion of livestock: what kind of animals farmers had, where they had come from, how they were kept, and what their values were to their owners. Sections from the second installment of this report that follow

---

49. A weekly newspaper published in Urbana, Ohio, 1838–41.

represent Affleck's interests and demonstrate what made his writing appeal to a regional audience. His reporting on animals, references to introductions of livestock into America, discussion of farm crops (in this case hemp), and descriptions of farms reveal Affleck's lifelong dedication to animal husbandry, his interest in regional agricultural endeavors, and his growing awareness of ornamental landscape features, present or absent. These articles foreshadow Affleck's efforts at the end of his life to rebuild livestock herds in Texas.

—Ed.

## Notes from Kentucky, by the Junior Editor [Extracts] [50]

At MR. ISAAC VAN MEETER'S we found what remains of the noted "Phyllis" now, alas! But a wreck—she has become diseased. She was, in her day, a most superior animal. A Hereford cow, and an old cow of the '17 stock, attracted me—but the fine heifer "Hannah More," pleased me better; as did also a handsome bull-calf.[51] We here entered that magnificent property owned by Capt. Cunningham and his son-in-law, Mr. Van Meeter. Here we have the blue-grass pastures of Kentucky in great beauty and perfection. Just think of nearly 3000 acres, clothed even at this season, with an abundant coat of green grass, in which you sink as in a snow wreath! The woods cleared of their undergrowth, and of all timber that is not really valuable— nothing left but fine, tall sticks of blue-ash, black walnut, etc.—divided into lots of 200 or 300 acres of gently undulating land, with a soil as rich and as deep as man could desire, and generally speaking well watered! Just think of all this, and of these noble pastures covered with such cattle as are only to be found in any great numbers, in the interior of this state—horses and mules in scores, all fat and in fine order, none of which are ever fed a mouthful of grain or fodder, unless when the ground is covered with snow! The fences good, with gates between each enclosure, well-made and well-hung, so that they open with ease and shut of themselves—an immense convenience when

50. From *Western Farmer and Gardener*, 312–22.

51. Herefords—a bull and two females—were brought to America from England in 1817 by statesman Henry Clay of Kentucky, where they attracted considerable attention and were introduced into local herds. Over time, offspring were absorbed into regional cattle populations, diluting (and eventually losing) the original characteristics of the forebearers. See http://www .ansi.okstate.edu/breeds/cattle/hereford, accessed June 12, 2012.

one is crossing the country on horseback. The men all friendly, hospitable, and jovial—and the ladies—but I must "leave the girls alone," or I shall never get along!

. . .

I now turn towards Frankfort, on my way home, and spent another day with Mr. R. W. Scott. This is one of the best ordered farms in the state. Mr. S. knows his business, and pursues it *as a business*. From a poor, worn-out badly arranged farm, he has in a few years produced a place well worthy of a visit. His fences are good, with rows of locusts planted in the corners—an old road that intersected the property, and all his *very* poor points are also set in locusts—swampy places drained—poor tracts manured, some of which prove the good effects of manure on hemp,[52] which I have heard doubted; all his fields improved in succession by clovering, etc., and all brought to good shape and size; and at each gate is a *painted number,* nailed in a conspicuous place, denoting the particular field; a fine fish-pond in his lawn, well stocked—a good garden-spot well laid off and fenced—the old orchard carefully pruned, scraped and manured—and a young one planted—his stock all in good order and well cared for—and, in short, the whole business of the farm carried on as it ought to be. His stock, the prime ones, are many of them descendants of his old '17 cow "Hetty," who has proved a mine of wealth to him. She is still alive, and now in calf by Constellation.

. . .

There seems to be a strange and unaccountable dislike, felt by the farmers in the West, to their wives and daughters displaying the slightest taste for gardening or the cultivation of flowers—and in Kentucky it is as evident as elsewhere. What a cheap gratification it is! Why deny it then? One hand,

52. Hemp was a major cash crop in the blue-grass region of Kentucky, where it was turned into cordage and bags for baling cotton in the Deep South. Following Affleck's report from Kentucky is an article (pp. 323–25) by Sands Olcott of Newport, Kentucky, on the preparation of hemp, discussing the author's methods of production and costs in an effort to "state in as concise a manner as possible, the process best adapted to the present cultivation and use of these articles in the United States." In response, Affleck gives his "hearty thanks" for Olcott's "sketch of his method and its results and advantages."

a few days each spring, would suffice to put the garden in order, and keep it so. The garden! What do we generally find it? A poor, miserable, badly fenced, weed-overrun smothered-to-death nook! We know some farmers, otherwise men of taste and judgment, who would absolutely eradicate their wife or daughters' little bed of flowers, to stick in their place a parcel of cabbage or onions, for which abundance of room could be found elsewhere— but then "*they* can see *no use* in such silly things as flowers—they would rather a precious sight, see a hill or potatoes or corn!" Shame, gentlemen, shame!—you can indulge in your Durhams and Berkshires, your race nags and your Bakewells[53]—your dog and your gun—in your cigar and your bottle of wine, and deny to those who are ever confined at home, the cheap enjoyment of a flower garden; most frequently cultivated, *you yourselves* know, that *your* homes may be rendered cheerful and pleasant *to you*. We have already remarked, that we found, to this, some honorable and praiseworthy exceptions.

T. A.

## TRAVELING INTO THE DEEP SOUTH

The following five accounts, from December 1841 and January 1842, given here in abbreviated form, describe Affleck's trip from Cincinnati to Baton Rouge, Vicksburg, Natchez, and New Orleans during which he exhibited and sold his own livestock, met with community leaders, farmers, and plantation owners, and attended regional agricultural fairs in Mississippi and Louisiana. The trip changed his life.

Affleck's stated purpose was to discuss, promote, and advance interest in agricultural issues and animal husbandry, but also we see in his writing a growing interest in landscapes. Sections relating solely to livestock (including names and characteristics of cattle, pigs, and sheep Affleck encountered) and detailed explanations of the results of an agricultural fair in Natchez have been excised in favor of those relating

53. "Bakewells" are sheep developed by British agriculturist Robert Bakewell (1725–95), recognized for his importance in English animal husbandry through contributions to the selective breeding of livestock, including sheep, cattle, and horses. While these appellations were common among farmers and in nineteenth-century agricultural journalism, they have now passed into obscurity.

to horticultural and agricultural topics, with summaries given where relevant. This trip to the Baton Rouge-Natchez-Vicksburg area proved significant in Affleck's life. He was well received by local residents, he felt comfortable there, and he saw opportunities—for both business and marriage—that would soon bring him back.[54] Affleck discusses towns he visited, the agricultural products—both plants and animals—he encountered, and the influential people he met.

While in the area, Affleck also visited the smaller communities of Washington (Mississippi) and Bayou Sara (Louisiana). Washington, about six miles northeast of Natchez, was the territorial capital from 1802 until 1822, when the capital was moved to Jackson. It was the home of Jefferson College, Mississippi's first institution of higher education, chartered in 1802 and named for Thomas Jefferson. This all-male school later became a military academy (second oldest after the United States Military Academy at West Point) that operated for over 150 years, closing finally in 1964. Jefferson Davis was a student there in 1818, and John James Audubon taught at Jefferson College in 1822–23. The "Fair" that Affleck describes here was sponsored by the Agricultural, Horticultural, and Botanical Society of Jefferson College, established in 1839. And it was to Washington that Affleck later moved upon his subsequent marriage.

Bayou Sara, Louisiana, is south of Natchez on the east bank of the Mississippi River. Founded in 1790 as a trading post, it was located near the Mississippi River below the bluff on which St. Francisville sits, a situation similar in many ways to Natchez-Under-the-Hill and Natchez. As a place where riverboats landed, Bayou Sara, like Natchez-Under-the-Hill (described by Affleck as "a spot celebrated . . . as the most perfect nest of iniquity extant"), was a community of colorful characters. Up until the early 1860s, it was the major shipping port between Natchez and New Orleans, but in 1862 it was burned by Union forces and never regained its former prominence.

Throughout his travels, Affleck was constantly investigating livestock, learning about local breeds, and exploring ways to take advantage of new markets, particularly for animals he owned. His discussions related to animal husbandry are summarized here.

Among the indigenous crops he discusses are the cushaw and cow pea, both of which were staples of the regional diet. Cushaw (*Cucurbita mixta*), a variety of winter squash grown in the Deep South, was domesticated by Native Americans centuries ago. The vine produces large leaves and spreads widely, quickly covering the ground.

54. Many residents of the rural areas that Affleck visited were of English and Scottish heritage, perhaps accounting for his affinity with this region and those who lived there.

A single vine produces numerous large fruits, averaging ten to twenty pounds and often growing to beyond eighteen inches long. The skin is whitish-green with mottled green stripes. The plant is heat tolerant, and the fruit can be stored for months, which made it an attractive and valuable crop for nineteenth-century farmers. Like other members of the squash family, the fruit has a mild flavor, and the seeds and flowers are edible. Today it is not grown commercially but is sometimes found in local farmers' markets. The cow pea, *Vigna unguiculata spp.*, is one of several species of the genus *Vigna*, widely cultivated throughout the world. Most common in the South is the black-eyed pea, *Vigna unguiculata* subsp. *unguiculata*.

On this trip, Affleck intended to see agricultural fairs in Baton Rouge and Washington, Mississippi. Travel difficulties delayed his arrival in Baton Rouge, but he did attend the fair in Washington at Jefferson College, and he gives a lengthy report of what he found. He noted the differences between fairs in Ohio and Kentucky ("somber assemblages of men . . . nothing to relieve the gloom of bearded faces and rail pens!") and here, how "the ladies had taken the thing in hand and . . . gave life and beauty to the scene." Later, Affleck discusses at length the Society's meeting, including its secretary's report, summarized here. It is worth noting that the Society's president was B. L. C. Wailes, one of the community's leading citizens.

Benjamin Leonard Covington Wailes (1797–1862), usually known as B. L. C. Wailes, exemplifies the economic, intellectual, and social life of the Natchez District in the first half of the nineteenth century. Born in Georgia, he moved with his family to the Natchez Territory in 1807. As a young man, Wailes was educated in classical studies at Jefferson College and entered the military, becoming a frontiersman in the region involved with land surveying and the negotiation of Indian treaties. He married in 1820 and established a plantation in Washington. He rose in rank in the military and served in the Mississippi legislature until it moved from Natchez to Jackson. Active in the Jefferson College Lyceum, he joined others in the region in educational, scientific, and agricultural initiatives, and though not a large planter, Wailes was among those who advocated for improved agricultural practices. In 1839 the Lyceum was reorganized as the Agricultural, Horticultural, and Botanical Society of Jefferson College. Initially its membership consisted of the trustees and faculty of Jefferson College, but it later accepted members chosen by that group, including most of the gentry of Adams County. Wailes served as president through 1843. As documented by Affleck's report, the society sponsored fairs with exhibitions of livestock, agricultural implements, vegetables, herbs, and flowers. Numerous committees were established, and they visited plantations to gather and publish information on crop production. Though the society eventually dissolved, it resulted in the formation of county agricultural societies throughout the state, including the Mississippi State Agricultural Society. Wailes was a promoter of this movement.

Wailes is perhaps best known for his interest in geology and paleontology, with a particular interest in fossils, reptiles (especially turtles), and American Indian relics. His private museum in Washington drew many visitors. He traveled widely over several states collecting fossils and studying geological formations. Few men of his time explored as many Indian mounds as did Wailes. He was involved in writing a detailed geological survey of Mississippi, finally submitting a report of almost four hundred pages, including maps and drawings, in 1854. It was printed later that year. He was also active in collecting and preserving historical documents (such as letters, plantation diaries, journals, and early newspapers), and he was an opponent of secession. Material that he compiled eventually appeared in 1880 as one of the first histories of Mississippi. His health declined prior to the Civil War, and he died in 1862.[55]

Affleck often records the names of people he met and provides summaries of his discussions with them. Examination of these names reveals that he often was talking to leading citizens of the communities he visited. For instance, in Natchez, Affleck mentions discussions with W. H. and Joseph Dunbar. There were at least three prominent but apparently unrelated families named Dunbar in nineteenth-century Natchez, into one of which Affleck later married.

Scotsman William Dunbar (1750–1810), a planter, explorer, and scientist, immigrated first to Philadelphia in 1771, then to the Baton Rouge area in 1773, then to Natchez, where he established a plantation in 1784. He married in 1785 and had nine children (four girls, then five boys; the eldest boy, born in 1793, was William Junior). He is today recognized by historians as one of the most successful planters, agricultural innovators, explorers, and scientists of the Mississippi Territory, although his contributions to America's early national years appear mostly in footnotes.

Like many eighteenth-century gentleman amateur scientists, Dunbar's interests included astronomy, botany, zoology, ethnology, and meteorology. Appointed surveyor general of the District of Natchez in 1798, he represented the Spanish government in determining the boundaries between Spanish and United States possessions in that area. Immediately thereafter, he became a United States citizen and began making the first meteorological observations in the Mississippi Valley. Dunbar attracted the attention of Thomas Jefferson, with whom he corresponded and who secured his admission to the American Philosophical Society. In 1804 President Jef-

55. See: http://mshistorynow.mdah.state.ms.us/articles/357/b-l-c-wailes-the-natchez-district -and-the-mississippi-historical-society, accessed June 19, 2012.

ferson commissioned Dunbar to explore the Ouachita River country, and in 1805 Dunbar was appointed to explore the Red River Valley. Among Dunbar's scientific concerns were investigations of Native American sign language, fossil mammoth bones, and plant and animal life.

In the Natchez District, Dunbar planted cotton, indigo, and tobacco, and he shipped ochre pigment to Boston. He also introduced the square cotton bale, invented the screwpress, and began extracting cottonseed oil almost a century before this process would be industrialized. On his main plantation, The Forest, he operated an observatory equipped with the latest European astronomical instruments. His particular concern was the observation of rainbows. One of his practical contributions was a method for finding longitude by a single observer, without knowledge of the time. His meteorological speculations included the theory that a region of calm exists within the vortex of a cyclone. Dunbar corresponded with American and European scientists. He also served as chief justice on the Court of Quarter Sessions and as a member of the Mississippi territorial legislature. His most important writing was the first topographical description of the little-known southwestern territory.

William Sr.'s extensive business interests and property holdings (including four plantations with slaves) were, upon his death in 1810, divided among his children, who ranged in age from four to twenty-four, and as they grew into maturity, they too became prominent in the agricultural, intellectual, and social life of Natchez. One might assume Dunbar's interests in scientific and agricultural matters were also inherited by his heirs.[56]

Affleck's article cites "W. H." (William Henry) in reference to the hogs he showed at the fair (summarized earlier) and "Joseph," noted for his interest in using cotton bagging instead of hemp for cotton bales. They were brothers, sons of another William Dunbar (1775–1826). William Henry owned six plantations in and around Natchez, and Joseph owned Arundo Plantation in Jefferson County. Both were active in the Agricultural, Horticultural, and Botanical Society of Jefferson College, and by their involvement, we can conclude that they were both prominent landowners in the area.

56. See "William Dunbar," *Encyclopedia of World Biography*, 2004, <http://www.encyclopedia.com>, accessed June 18, 2012. See also http://mdah.state.ms.us/manuscripts/z2147.html; William Dunbar Jr. and Family Papers, Mississippi Department of Archives and History (Z 2147.000 S) ca. 1791–1984, accessed June 18, 2012. A recent biography is Arthur H. De Rosier Jr., *William Dunbar: Scientific Pioneer of the Old Southwest* (Lexington: University of Kentucky Press, 2007). An earlier account of Dunbar's accomplishments appears in Franklin C. Riley, "Sir William Dunbar: The Pioneer Scientist of Mississippi," *Publications of the Mississippi Historical Society* 2 (1899): 85-111.

At the time of Affleck's visit in 1841, Natchez was a community of fewer than twenty thousand residents, many of whom were slaves. Affleck certainly would have observed slave conditions firsthand, and when he writes about these conditions, he does so in some detail. Through activities of the Agricultural, Horticultural, and Botanical Society, Affleck met prominent community leaders with whom he shared a Scottish heritage and interests in the natural sciences and agriculture. During this visit, Affleck met Anna Dunbar Smith, the widowed daughter of Isaac Dunbar (1778–1849), a wealthy planter and landowner. Isaac was one of nine children of Scotsman Robert Dunbar (1748–1826).

Altogether, Affleck's trip to the Natchez area in late 1841 was a significant turning point in his personal life and professional career. He subsequently moved to Washington, married Anna in 1842, took over the management of her financially distressed plantations, and started what became one of the first commercial nurseries in the South. He continued to write and soon developed business contacts in New Orleans and new markets for his nursery products and his writing. It must have seemed that the family tragedies and business failures of the past were behind him, and the future looked promising.

—Ed.

### A Trip to the South [Extracts] [57]
By the editor.
Washington, Miss, Dec. 2, 1841

TO MR. C. FOSTER [another editor of the *Western Farmer*],

MY DEAR SIR—I may say, with all safety, that since I left Cincinnati, nearly a month ago, this is the first leisure hour I have had. I will use it by giving to the readers of the Farmer and yourself a sketch of my progress so far.

As you know, the object of my trip has been partly to make sale of what little stock I brought with me, but more, by scattering it over the country, to stir up and nourish a taste for the improvement of the common stock, and the introduction of new and improved varieties, and still more, to see, learn and report the state of things amongst the planters of the South, the farming community of this much belied region.

I left Cincinnati in abundance of time, as I thought, for the Fair at Baton

57. From *Western Farmer and Gardener*, 513–20.

Rouge, which I had a most anxious desire to attend. But one delay occurred after another—the river was so low that we got frequently aground during the day, and could not run any during the night—that I did not reach Vicksburg, still a good day's run above Baton Rouge, until the morning of the last day of the Fair; and feeling vexed, disheartened and disappointed, and the stock being by no means in a thriving state on board the boat; and having, moreover, a good many of them to leave there, I resolved to land all, and defer my visit to Baton Rouge until after the Fair at this place (Washington). I have since learned that I was looked for, and that the non-arrival of the stock caused some disappointment—that the fair went off well, though the show of stock, implements, products, &c., was by no means as good as could have been desired. Being the first attempt at an exhibition of the kind in the State, little else could be expected; but I feel confident that the next effort will prove much more successful. But I must retrace my steps.

I found, on landing at Vicksburg, that I had been looked for there; and within twenty-four hours after my arrival, I had some twenty or thirty of the most spirited and influential planters of the vicinity in town to see the stock. I very soon found that I should have to limit my sales; for having only some fifty head of thorough-bred hogs, and ten head of choice Durhams, headed by "Cincinnatus," and having it in view to leave a part at this place, a part at Baton Rouge, and a few at Bayou Sara, I could not allow all to be carried off here. Within three days, I made the following sales: . . .[58]

I had very little expectation of finding so much of a spirit of improvement excited here. All seem alive to the advantage of improving their stock, and especially their wild swamp hogs. Though the fine Durhams I had along attracted a good deal of attention, there seemed to be a strong dread of investing money in them; several gentlemen in this vicinity had introduced cattle from the north (Ohio and Kentucky,) and with but very partial success— some losing all, others only a part; and some again saving all. . . . [*Affleck then discusses conditions among local herds of livestock and local leaders in improving livestock herds.—Ed.*]

I remained in Vicksburg—a singular, scattering town, built on a series of high knobs, and though by no means a pleasant looking place, presenting

---

58. Here Affleck lists names of men who bought his stock and discusses their attributes. Many of the names he mentions, such as jurist W. A. Lake, businessmen W. H. and A. M. Paxton, and Dr. George Smith, were prominent citizens of Vicksburg.

a very gay and beautiful appearance from the river at some miles below. I remained here about a week, making the acquaintance [of] almost all who felt any interest in stock matters. I made only one short trip into the country; that as I mean to repeat the same trip, more at my leisure, I will pass it over for the present.

Leaving all the cattle but five, and with only some ten or a dozen of the hogs, I came down to Natchez, two days previous to the Fair at this place. I found this city very beautifully situated, and a neat, pleasant looking place. The town proper is on a high plain, on top of a perpendicular bluff, at the foot of which, on a narrow and irregular strip, is scattered "Natchez under the hill,"—a spot celebrated, until of late years, as the most perfect nest of iniquity extant. Natchez is recovering wonderfully from the effects of the dreadful hurricane, which left it almost an entire pile of ruins, in May, 1840. The business done here is very great—and certainly prices of everything are high enough to warrant fortunes being made in a year or two. . . . [*Affleck complains of the high cost of hiring locals to transport goods in Natchez; then he remarks on how his Durham "Cincinnatus" attracted attention. He didn't sell any of his cattle since he had promised to exhibit at the Washington fair; instead he loaded them on a train for shipment to the fair, noting how "cold and blustering" the weather had turned.—Ed.*]

You know the plain, and matter-of-fact business-like manner in which the Fairs in Ohio and Kentucky have been held? A somber assemblage of men, ambitious only of each exhibiting the finest animal in its class, or the best machine or implement of its kind—nothing to relieve the gloom of bearded faces and railpens! Here, it was altogether different—the ladies had taken the thing in hand, and not only gave life and beauty to the scene of their own presence, to the number of several hundreds of as lovely forms and faces as eye need wish to rest on, but had on the previous day hung every port and pillar with garlands of flowers, and covered the walls of the rooms, in which the fruits, &c. were exhibited, and the porch, with wreaths and bouquets. When I tell you that the rose is still in bloom, and that the gardens are everywhere stocked with splendid evergreens, which with us require the shelter of a hot house, you will understand the more readily how this was possible, so late in the season. The vegetables and implements were laid out under a cluster of oak trees, still in full leaf. All went off well—so well that I would willingly particularize to some extent, but must restrain myself. Knowing as you do, how I dislike this pretended extra degree of refinement in the wives and daughters of our farmers, that leads them to make such fools of themselves as to be *quite shocked* at the thought that *they* should know a horse from a cow, or a hog from a sheep, you can appreciate my gratification at seeing the daughters of the polished South, moving leisurely about amongst the stock, even stopping to admire the beauty and docility of "Cincinnatus," and venturing to lay hands on Mr. Hall's Colossus! But to the show. As well deserving the precedence, which in point of fact he maintained, I must speak of the stock, &c., shown by Mr. Thomas Hall, of Adams county.

. . .

*[Then follows a discussion of Mr. Hall's display of "two genuine Mississippi productions—a fine young bear cub from the cane brakes, and a monstrous live rattlesnake, selected from a fine breed that Mr. Hall keeps upon his plantation to catch rats and trail abolitionists!"—and hogs, giving evidence that "Mississippi could raise her own pork." Affleck also discusses cattle and sheep, noting his disappointment in what he saw. The horses, however, "carried the day. Some beauties were shown, evincing the taste felt here in that species of stock."—Ed.]*

Pumpkins, squashes, cushaws, &c., were shown in infinite variety, and almost all mammoth. The turnips were the finest I have ever seen in this country. The planter of the South cannot too highly appreciate his crops of pumpkins and of turnips, as fall and winter feed for his stock, and the sweet potatoes are a mine of wealth to him who wishes to make his own pork. The crops of them grown in the South are immense. Hr. Hall alone has housed some 2500 bushels this year; and Col. P. Harrison, in the same neighborhood, about 1000; the same gentleman has on hand about 3000 bushels of old corn! Hundreds of bushels are left in the ground to be picked up by the hogs. I shall have much to say by and by upon the renovation of exhausted cotton farms here, by feeding off such crops. The cow pea too is invaluable.

Specimens of corn, and good ones too, of cotton; one variety of which grown by a planter of this county, and said to be a cross of the Egyptian on the okra or twin cotton, was to me a beautiful and interesting object; it is represented as yielding the enormous crop of 3000 lbs. (in the seed) to the acre—of tobacco, grown here from Havana seed, and manufactured on the plantation of Peter Rucker, Esq., was a superior article; English walnuts, grown by Mr. John Robson, who also has brown almonds in great perfection; a tempting lot of green peas, from Dr. Metcalf's garden; a most beautifully put up bale of cotton, from the plantation of Joseph Dunbar, Esq., packed in *cotton* bagging, and bound with hoop iron, and so well compressed as not to require the application of the steam press at New Orleans. Mr. Dunbar was, I believe, the first planter in this region to rebel against the glaring impositions practiced upon them by the manufacturers of bagging and bale rope from hemp; his different attempts at finding a substitute, I shall speak of again; as also to my Kentucky friends in what manner and to what extent their brethren here have been imposed upon, and the hemp-growers' interests so seriously injured by the malpractices of manufacturers. Mr. Dunbar has promised me a statement of *facts* coming under his own eye.[59]

Some plows and other implements were shown, but there was rather a dearth of such things. How little the different manufacturers study their own interests, in overlooking the advantages they would derive from exhibiting their machines and implements at these fairs!

The stock I brought down and had on the ground, consisting of Cincinnatus, Coronet (by Carcess, dam imported Princess) and other beautiful heifers, all by Dennis' Coronet, attracted a large share of attention. They so evidently required some improvement to be made upon their stock of cattle, that I was very anxious to have some of those here. I proposed the formation of a club for the purchase of the white bull, which was at once taken in hand by the spirited president of the society, and in a few minutes made up. The price I named—and a low one for such an animal—was $600—which was taken as follows, in shares of $25 each. . . . [*A listing of names is given of those who bought shares in the animal.—Ed.*]

But to the fair again. Not the least comfortable part of the arrangements consisted of the splendid lunch-dinner, in fact, set out, the materials for which were furnished by the members of the society. And certainly a most luscious affair it was. Several bacon hams, raised, fed and cured in Mississippi, would vie with the finest put up in the queen city—but the dish of all others of which this country may be proud, is her mutton; several saddles

59. As previously mentioned, there was frequent discussion in agricultural papers concerning the use of hemp, grown in blue-grass regions of Kentucky, and its application for baling cotton, grown elsewhere in the South.

were here, such as would vie in every respect, with the delicious mutton of my own native hills—to these I paid especial attention! The rich jellies, and still richer and more to be prized butter—the salad, &c., I shall pass over for the present.

The committee on plantations awarded the premium for the best improved and best cultivated farm, to Jos. Dunbar, for his farm Arundo.[60] The report is an interesting one, but I afterwards visited the farm myself. I shall not quote it now. I send you the secretary's published report.

. . .

---

*[A lengthy summary of the society's meeting follows, in which resolutions were approved (including one that approved distribution of twenty copies of their report to the* Western Farmer and Gardener*) and awards ("premiums") are listed for livestock (including horses, mules, cows, pigs, sheep); vegetables (corn, squash, pumpkins, "Mississippi Tobacco," and cotton); plum jelly ("To Mrs. A. M. Smith"); and farm equipment. The "committee" also recommended that a copy of the* Western Farmer and Gardener *be given to twenty "who have been the largest contributors to the former exhibitions of the society, and have received the greatest number of certificates." Elections for new officers were held, and a resolution was passed: "That the* Western Farmer and Gardener*, a monthly periodical, edited by Thomas Affleck, Esq., corresponding member of the society, be recommended to the patronage of the members and the community at large, as one of the most valuable agricultural works in the United States." The society then adjourned, to reconvene a few days later (Saturday, November 27, 1841), when it unanimously "resolved" that "the certificate of the society be awarded to Thomas Affleck, for a valuable importation of Durham cattle and Berkshire hogs into the State," which Affleck "handsomely acknowledged." New members were elected and the society adjourned until its next meeting on the last Friday in April. The report is signed by B. L. C. Wailes, President.—Ed.]*

---

To say that I *felt* the honor done us by the society, would be to say but little. That a body of such men as compose this society should give such a gratifying proof of the estimation in which they held the result of our labors in the noble cause of agricultural improvement, was indeed encouraging.

60. Arundo was a plantation of 1,200 acres located in Jefferson County, owned by Joseph and Olivia Dunbar, mentioned above. No records or images are known to exist of the plantation.

The certificate awarded me is a very neat one, and shall grace the walls of our office, in a frame worthy of it. You will understand that I exhibited nothing for premiums, or in competition with the members of the society.

I must now close the account of my trip for the present, but will continue it in the next number.

THOMAS AFFLECK.

---------◆---------

## A Trip to the South [Part 2] [61]
### Washington, Miss.

MY DEAR SIR,—I believe I wound up my last communication, by giving the result of my observations during a visit to the finely conducted plantation of Mr. Joseph Dunbar. On the morning on which we left his hospitable mansion (and by the way, it seems I there missed a treat in not seeing Mrs. D.'s admirable dairy arrangement. I knew nothing of it until afterwards; but will assuredly have a peep the next time I wend that way!) we all proceeded to that of Mr. Thomas Hall, the gentleman who made such an exhibit at the fair. The house and homestead are placed upon a poor oak ridge; the plantation being at some distance, and lying along Coles' Creek. We again took a careful look over all his improved stock—the large bull Beltzhoover; his cows; "Tom Hall" (!!) and "Minor" hogs; his pair of Berkshires &c. I was highly gratified with the garden. Here I found an immense *patch* of cabbages, many of which would have done credit to the best Dutch garden about Cincinnati; fine hedges of an evergreen, of which I shall speak again; celery, turnips, lettuce, &c., &c., in great perfection, and this too, on a piece of land originally considered almost worthless. Here, too, I saw some strips of blue grass, and of red clover, looking as vigorous and green, and of as good a growth as any in Kentucky—but Mr. H. seems to think that neither will do well unless cultivated; the extreme hot, dry weather of the summers here kills both. I cannot but think that if proper locations on the farm were selected, and a proper course pursued in the sowing and after care, that both would succeed. Mr. H. ascribes the fine health enjoyed by his negroes to the abundant supply of rain water contained in his numerous cisterns.

---

61. From *Western Farmer and Gardener*, 543–45. The "last communication" to which Affleck refers does not appear in the compilation of articles published as a book.

My next visit was to the farm of my very kind friend, Mr. J. W. Bryant, within one mile of this place. It is here *Cincinnatus* has taken up his abode, and most excellent quarters he has. Mr. Bryant has been improving his stock for many years, and has some pretty fair animals. His little flock of native sheep—the old long-legged, naked bellied Spanish sheep—I admired very much. Faulty as their form is, in some respects, and worthless as their coat most generally is, their remarkably heavy, full hind quarters, and the delicious mutton they afford, prevent their being passed by with indifference. After looking over the flocks and herds; the half acre of Gama grass, which Mr. B thinks highly of; Mrs. Bryant's well stocked garden, which her taste for flowers has rendered a complete *omnium gatherum* [miscellaneous assortment]—we adjourned to the Ginhouse, which has, throughout my trip been a favorite resort with me.[62] Here I saw the only cotton-thrasher I have yet met with, though it should be in every ginhouse in the South. It is intended to clean out the dust and trash from the seed-cotton, previous to its being ginned, and in this succeeds admirably. I have made a rough sketch of it sufficient for you to engrave from. As a thing that will be interesting to our Northern and Western friends, I procured samples of the cotton in all its stages.

Mr. W. J. Ferguson, close by Washington is a keen Horticulturist and occupies much of his time in that way. He has also been procuring some Berkshires, and makes *strong threats* of going on until he has a fine herd.

Of Col. Wailes I have already spoken. He has for many years been experimenting with fruits, flowers and vegetables, in a scientific manner; and I am in hopes of procuring from him, the results of his experiments. It is impossible to converse long with him without acquiring a fund of information upon such subjects; still I am anxious to have it from his own pen. He has promised to forward to you various rarities—such as the Cherokee rose, live

62. A ginhouse is a large, barnlike structure where cotton was ginned, or prepared for market. The term *gin* is derived from the "cotton engine" invented by Eli Whitney in 1793. This mechanical process involved separating the fibers, called lint, from the seeds and usually involved using machinery powered by mules. The lint was moved into another area and packaged, using mules to pack the lint into loose bales weighing between 400 and 500 pounds. These bales were shipped from the plantations by oxcart, raft, or steamboat to ports such as Galveston and New Orleans, where they were compressed into smaller bales (by a hydraulic process, called a "compress") for delivery to mills in New England or England.

oak, varieties of figs, grapes, &c.; some sods of Burmuda [*sic*] grass, Spanish clover, Texan rye, &c.

The ladies of Washington and the neighborhood are enthusiastic and successful gardeners—particularly in the floral department, though the kitchen and fruit gardens are by no means neglected. I propose taking a ride along Second Creek, where I am told I shall find several extensive and well stocked Greenhouses.

Col. Knight, some four years ago, purchased what was then considered, a completely worked out cotton plantation, with the view of making a dairy-farm of it. Its surface was just sufficiently rolling to allow every particle of soil to wash away, and to form innumerable gullies. He has gone to work on this unpromising tract, knowing that the sub-soil was good, and would form the basis of an excellent soil. After a world of labor in plowing down the ridges between the washes, and so rendering the surface tolerably level, he has, by sowing and feeding off cow-peas, rye, oats, turnips, &c., rendered a great part of his farm almost as good as ever it was. In this he has derived great assistance from his little flock of sheep, the remnant of that originated and kept up for some years by Col. Wailes, being a cross of the Saxony on the common, old Spanish sheep. It is to sheep that Mississippi must look, to afford another and a more valuable staple than cotton, and to reinstate her exhausted cotton farms. Col. Knight showed me a small Bermuda grass meadow, from which he cut a crop of excellent hay last summer, and which yielded the extraordinary quantity of nearly 4 tons to the acre!

I spent a day with Col. Claiborne, at Martha's vineyard.[63] The Col.'s zeal in the cause of agricultural improvement is great; and as editor of one of the most widely circulated papers in the State, he has it in his power to aid

63. The reference here is to Colonel Ferdinand Leigh Claiborne Jr. (1809–93), son of Ferdinand Leigh Claiborne (1772–1815) and nephew of Governor William C. C. Claiborne (1772–1817), the second governor (1801–1803) of the Mississippi Territory. The elder Ferdinand moved to Adams County in 1802, and both were prominent in the military, territorial matters, and Indian affairs. While William was governor, the territorial legislature created Jefferson College in Washington. William left Natchez in late 1803 upon instructions from President Jefferson and became the first governor of the newly acquired Louisiana Territory. Members of the extended Claiborne family were landowners and prominent in the community. From the eighteenth century onward, members of the Claiborne family have served as governors and members of the U.S. Congress, particularly in the House of Representatives. Descendants of the Claiborne family continue to be prominent in local and national politics and government.

the cause immensely—and ably does he use that power. The farm on which he now lives, is an old place much washed and worn; but though only in his possession for a single year, he has made great improvements. He has adopted Sir Henry Stewart's [sic] views, in planting shade trees, and carried them out in a manner that will tell, in the added beauty and comfort of Martha's vineyard, in a year or two.[64] Fields of rye and of turnips—the latter being the best variety cultivated, Dale's Hybrid, and a fine crop—are proofs of his management. He is improving the stock on the place rapidly, having amongst others a right good imported cow, of the short horn breed. By the way, Col., don't forget to send me the recipe for cooking that dish of squash! It was too great a luxury not to have it in common use.

As I am just about setting out for a bear hunt, I will close for the present—and

Am yours, &c.

---

*[Although unsigned, this article and the two that follow were clearly written by Affleck.—Ed.]*

---

———◆———

## A Bear Hunt in the Big Brake [65]
### *Natchez,* Dec. 15, 1841

A bear hunt! aye, and not much bear hunt after all. Still such as it was, you shall have it, with an account of all I saw on my first ramble through the Cane and Palmetto brakes of Louisiana.

Mr. B. Alderson, of Washington, is opening a plantation in the big brake, in Concordia Parish, some five and twenty miles from Natchez. All that region of country has been entirely unsettled until within a year or two, hav-

64. Affleck's reference here is to Sir Henry Steurart (1759–1836) of Allanton House, Scotland, and his well-received work, *The Planters Guide: Or a Practical Essay on the Best Methode of Giving Immediate Effect to Wood by the Removal of Large Trees and Underwood, etc.* (London, 1828). An American edition appeared in 1832. Largely unknown today, Steurart in his day was an influential agriculturist in Scotland, and his work on trees was widely circulated in England, Scotland, and America. See A. A. Tait, "The Instant Landscape of Sir Henry Steurart," *Burlington Magazine* 118, no. 874 (January 1976): 14–23.

65. From *Western Farmer and Gardener,* 545–47. The American black bear (*Ursus americanus*) is native to Louisiana; a brake is a shrubby thicket of cane or palmetto.

ing been so much under water during the flood of 1828, as to debar almost all from making the attempt. Some few of the veriest [*sic*] wilderness-loving squatters only had set themselves down, and made a precarious living by hunting; and it was the success of a party of these that induced my friend Mr. A. to propose to me to go and spend a week with him at his cabin in the brake, and join these hunters in a regular bear hunt. Accordingly, shouldering my fowling piece, and mounting an old brute of a hired horse, I joined him and we set out.

Crossing the river to Vidalia, a scattering little town, the county seat of Concordia; and the locale of friend Smith's "Concordia Intelligencer," one of the most pleasing and spirited papers I know of.[66] We rode for some miles through old plantations, some of them most carelessly, and other most admirably managed, along the shores of the lake from which the parish takes its name. This is a beautiful sheet of water, and like the river, presented to me, the singular fact of being leveed or banked in, to prevent its overflowing the country. This embankment and the shores of the lake for many yards are completely covered with a solid sod of Bermuda grass, and have a very pretty appearance. The day was pleasant, the lake placid and beautiful; the road as fine as you could wish to drive over; the character of the country new, and of exceeding interest to me; large gangs of laughing, singing, happy looking, well clad negroes at work in every field; my companion, a pleasant one—so that altogether with my anticipations of sport at our journey's end, and a little amongst the ducks as we went along, I do not know when I have had a more agreeable ride. After leaving Lake Concordia, we entered the Cane, when the scene changed; and by the time we reached the plantation of Mr. Lillard, with whom we sojourned for that night, I was heartily tired of it.

Mr. L. has settled partly upon land not yet in market, and consequently claims, under the pre-emption law.[67] It is only some two or three years since he commenced operations, but having a strong force, and being moreover

66. C. S. Smith was an editor of the Concordia *Intelligencer,* published from 1841 through the 1880s.

67. The Preemption Act of 1841 (27th Congress, Ch. 16, 5 Stat. 453), designed to "appropriate the proceeds of the sales of public lands . . . and to grant 'pre-emption rights' to individuals" (known as "squatters") already living on federal lands, greatly benefitted territorial settlers in the 1840s and 1850s. Under this act, states, including Mississippi and Louisiana, would receive 10 percent of the proceeds of the sale of public land, and those already living on lands owned by the federal government could purchase up to 160 acres at a low price before the land was

his own manager, and an efficient one, he has a large and fine plantation now open. Though rich as it is, it is too low, and too near the level of the swamp for my taste. My observations on the excellent condition of the negroes will come in their proper place.

Soon after leaving Mr. L.'s we entered the cane-brake in earnest. Think of those long, 20 or 30 feet canes, that you see occasionally used as fishing rods, standing over the whole face of the country, so think that it would puzzle a good dog to find its way through them; and clothed towards the top with rich green leaves! Through this the path is cut, the cane not infrequently meeting overhead. It is occasionally varied by large palmetto brakes, the tops of the plants reaching up to one's feet or even knees, on horseback. The cane land is some *few inches*, say a foot or two higher than the palmetto land, and hence more highly valued. It is of a lighter and more sandy nature, and as rich as I suppose it is possible for land to be.

About noon we reached Mr. Alderson's cabins—but I must cut short my rambling epistle, and pass over our calls at the various plantations in the Brake, Messrs. Huddleston's, Johnson's, Glasgow's, and Col. Thompson's, and come at once to our appointed meeting with the Bear-hunters. There are some five or six of them who hunt in company, and have kept Natchez fully supplied with Bear meat this winter. Immediately back of Mr. Alderson's cabins they had recently killed some half dozen, and here accordingly, we

---

offered for sale to the general public. There were eligibility requirements (head of household; single male over twenty-one years of age; U.S. citizen; resident of the land for at least fourteen months); the claimant had to improve the land for at least five years; and if the land remained idle for six months the government could reclaim it. In addition, the act granted 500,000 acres of government land to each state provided that the proceeds from sales were used for "internal improvement . . . namely roads, railways, bridges, canals and improvement of water-courses . . . and draining of swamps. . . ." This legislation later became the basis for the Morrill Act of 1862, which again allocated federal property (30,000 acres per Congressional representative based on the 1860 census) to the states, excluding those "in a condition of rebellion or insurrection against the government of the United States," provided the proceeds of sales were used to establish institutions of higher education in military, agricultural, and mechanical arts; the act was amended in 1890 to include states of the former Confederacy. This act resulted in the creation of what became known as "land-grant colleges," which provided agricultural education to the general public, a cause that Affleck and his contemporaries had strongly championed throughout their careers. Kansas State University, opened in 1863, was the first land-grant institution created under the act. Many schools of landscape architecture later emerged from these public land-grant colleges.

proposed joining them. We had an early breakfast with Col. Thompson, and riding over to Mr. A.'s found Gray, one of the hunters, impatiently awaiting us, with the pack just coming in hearing, and in full cry.

Shouldering his rifle, friend Alderson dashed into the cane on foot; whilst Gray and myself on horseback, galloped on to a point where we hoped to intercept Monsieur Bruin. Occasionally stopping in our gallop we heard the dogs pressing on; whilst Gray called my attention his favorite dog—"hear her—youp, youp,—that her,—all's right—follow on." The dogs used are a sort of cur, with a dash of hound dog, and were certainly, though they have proved themselves otherwise, a most unpromising *looking* pack—some of them most woefully cut and scarred. On we pushed until our road came to a termination, when we abandoned our horses and took it on foot. Then came the toil of a bear hunt in the cane brake! But with the music becoming clearer every moment, who could think of that? The favorite dog still led on—all was right—"wherever she is, there is the Bear"—at one time I could hear the awkward brute scuffling along within a few yards of us, but totally invisible—this only excited me to greater exertions, but in vain; for after a close chase of five or six miles, the dogs drew off and returned to us, and we had to give it up, very loath.

My companion remarked more than once, that we had only a part of the pack, and that two bears must have been started. And on our return to Mr. Alderson's cabins we found him and another of the hunters, and by them lay a dead bear—a fine fat cub of about 100 pounds. His history of the chase ran thus. He had no sooner entered the cane, than he found that the dogs were rapidly approaching him, and in a few minutes, he saw them and *the varmint* within a short distance. He got into a thick briar patch, where our friend left his coat-tails! Here the dogs became somewhat loving, and his bearship, not to be out-done, turned and took the leader in his warm embrace, at the same time muzzling in a most affectionate manner about her throat. *Lady* begged in such a pressing manner to be let off for this time, that friend A. rushed up to insist upon it; and taking a quiet aim at the whereabouts of bruin's heart, he pulled trigger,—and alas! Alas! Some instantaneous change in the position of the struggling animals, brought the slut's head within range, and the ball [bullet] made double work of it, passing at the same time through her brain and the bear's heart! Here was an unfortunate affair! The best dog in the pack! But well was it for him, that his ball told so fatally to the bear, for so completely was he entangled in the briars, that had his wound been a slight one, he would have had no recourse but his knife, and the rencountre, in all probability would have been a dangerous one. He was shortly joined by one of the hunters, by whose aid the game was brought in. When Gray was informed of the death of his favorite, I felt for the poor fellow. Only those who have loved and followed a favorite dog for years, can tell what he suffered. "And what will the old woman and the children say? Poor poor *Lade*!" Alas! Poor Lady!

We now turned our face homewards, and again spent the night with Mr. Lillard, enjoying his hospitality to the full. For you must know that Mr. Alderson has just commenced operations; and rarely visiting the swamp himself, and still more rarely entertaining his friends there, the choice of the softest plank, was *almost* all he could offer! On our way to Mr. L.'s I shot down a most excessively fat 'possum, and we carried him along. The fellow had been making so free with the fine, riper persimmons, with which the brakes abounded, that he could scarcely waddle.

Next morning we breakfasted off the gentleman, cooked in *old Varginny* [Virginia] *style*—baked with sweet potatoes, and served with pumpkin sauce. Ah! my dear sir, you must visit the swamp and the wilderness to eat game in perfection.

Yours, &c.

---

### Letter to Charles Foster [68]
### New Orleans, Dec. 27, 1841

To Charles Foster, Esq:

*My dear sir,*—Here I am, at that noble structure, the St. Charles Hotel— certainly the most comfortable, homelike yet luxurious establishment within my *ken*. I have seen but little yet of New Orleans. Soon after my return from

68. From *Western Farmer and Gardener*, 547–48. The St. Charles Hotel held an important place in the community, as described in John Kendall's *History of New Orleans* (Chicago: Lewis Publishing Co., 1922), 685–88: "The [first] St. Charles was the first large building erected above Canal Street. From the day when its foundations were laid down to the close of the century, when its supremacy was successfully attacked by the construction of other large and luxurious hostelries, it was the representative building of New Orleans. It shared the fortunes of the city, good and bad; it prospered when it prospered, it suffered when it suffered. Within its walls half the business of the city was transacted over a period of fifty years; and there for a still longer

the swamps of Louisiana, I learnt that a lot of Berkshires, you had shipped to me, had arrived in Baton Rouge, I took the first boat for that place, and on my arrival found a very pretty lot of pigs, under the care of friend Huston, Editor of the Gazett[e][69]—an enthusiastic lover of fine stock. I spent a few days with him, visiting some of the sugar plantations, and enlightening myself on the subject of sugar making.—Unfortunately the season was nearly over, so that I could not satisfy myself entirely; we spent a day with F. D. Conrad, Esq., who has been, and continues to be one of the mainsprings of the movements in favor of agricultural improvements in this State.[70] He has been doing some little in the way of improving his stock; but neither here nor at Natchez, do they go rightly to work. A part-bred animal, of either hog or cattle kind, provided it can be had for a few dollars less, is preferred to the thorough-bred. This will soon work its own cure. Every town and plantation along the river has been flooded this winter with part-bred cattle and hogs,

---

time half the history of the State of Louisiana was written. . . . The St. Charles was the first of the great American hotels, and it won for the city the reputation of being the most enterprising, as it was already credited with being the most aristocratic and possibly the wealthiest city in the country. It had a magical effect upon the quarter of the city in which it stood. It rapidly built up the First District. . . . St. Charles Street, which did not extend far above the hotel, was at that time the gayest and most animated thoroughfare in the United States, and possibly in the world. Between Lafayette and Canal streets it exhibited an almost continuous line of bar-rooms and restaurants. . . . Many [people] were attracted to the wonderfully prosperous city as a place in which to make a fortune rapidly. Here they remained six months or less at a time, and then fled northward or to Europe for rest and recuperation, before returning for the winter's strenuous labors. This was the element to which the hotels and restaurants catered. It was the custom to lodge at the hotels, but to eat at one or another of the countless restaurants which lined the thoroughfares opening into Canal. . . . It is said that several hundred outsiders dined every day at the St. Charles." After the original St. Charles burned to the ground in 1851, a second hotel, equally as grand, opened in 1852. It too burned to the ground in 1894, and a third hotel was subsequently erected but was demolished in 1974, fueling a major push for architectural preservation measures later enacted. Frederick Law Olmsted stayed at the St. Charles Hotel in the early 1850s while in New Orleans. About this time, Affleck concocted a scheme whereby he would send garden flowers grown at his nursery in Washington in a compartment on a riverboat, to be collected at the dock and transported to the hotel on a custom-made cart, to be sold to patrons of the hotel. The venture was not a success.

69. The Baton Rouge *Weekly Gazette* was published from 1819 to 1856 (in French and English) and continued to 1865 as the *Weekly Gazette and Comet*.

70. Conrad was a prominent lawyer and plantation-owner in East Baton Rouge Parish, active in initiatives to improve agriculture in Louisiana.

sold for thorough-breeds, when practicable, and then of course at three or four times the real value; but occasionally sold for what they really are, and then at double what they are actually worth. Purchasers should recollect with what ease they can create half and three-quarter bred animals; and how doubtful is the improvement effected by breeding to a part bred male.

Finding that it would require *time* to do anything here; and that that time spent in New Orleans would be sufficient for me to complete my business there, I came down—and I have just learnt that Mr. Huston had removed all the hogs into the yard behind his office, which being in the very centre of the burnt district (of the serious fire that took place in Baton Rouge on the 23d inst.) I must fear that all, or part of at least, of the unfortunate animals have been destroyed. Mr. H. has suffered with a host of others a very serious loss, in presses, types, paper, &c.

The banks of the Mississippi—the coast, as they are termed, present a very peculiar appearance. The plantations are strung along without inter-mission; and with the handsome, singularly planned mansions, the large sugar houses, the strings of negro's quarters, with the dwellings of the over-seers, form altogether a life-like scene. Some few clumps of live oaks, and groves of oranges are visible; but generally speaking the whole country is so much below the level of the river, that the *levee* prevents much of the state of things about the premises, being seen.

From what I have told you of the state of the weather and of my health, you will not expect me to satisfy much of your curiosity about this city. I meant to have seen all the gardens and green-houses, &c., but found it im-possible. I looked in upon one small scale green-house, that of a Mr. Batch, and found it in fair order. I took a peep into the seed stores, and found in one of them a right canny countryman of my own, Mr. Wm. Dinn.[71] He seems made for the business—careful, attentive, and cautious and has, during the number of years he has carried it on, established for himself an enviable reputation. He imports direct, those seeds that it is necessary to import, together with bulbs, tools, books &c., and is altogether doing a very flourishing business.

The markets always afford me a very pleasant and interesting stroll, when in a large city. That of New Orleans exhibits the most complete *mélange,* both of *marketables* and of people;[72] of any city in the world, I presume. French is the language of which you *hear* most. Tropical fruits are piled upon the same stand with Irish potatoes and onions—oranges, bananas, and pine apples are here in immense quantity. Innumerable varieties of the field pea of the south, and of lentils; of hominy &c., are sold. I procured specimens of almost all.

I examined the market closely. I had heard it highly spoken of, both as to variety and quality; and it *is* well supplied; but to compare it with that of Cincinnati is absurd. One Englishman, a butcher, with whom I conversed, and who saw me eyeing a well fatted beef that hung in his stall, side by side with some splendid hind quarters of mutton, and seeing I was from

71. Advertisements for William Dinn's "Seed and Flower Root Store," 11 Common Street at the corner of Tchoupitoulas Street, appeared in the *Daily Picayune* between 1837 and 1846. He advertised seeds for garden and farm (from the "Northern States, and from the first seed establishments in France, Holland, England and Scotland . . . fresh and genuine"), agricultural tools, and birds (including canaries, goldfinches, mockingbirds, and "Other songbirds"). Affleck had a business association and friendship with Dinn that lasted throughout Affleck's life.

72. There were many municipal markets throughout the community, but Affleck is probably referring to the French Market at the lower end of the Vieux Carré, near the Mississippi River. Originally a trading area for Native Americans, this market's first permanent structure was built in 1782. The market complex here was built in 1813, and by 1835, spaces were occupied by vendors of Native American, African American, European, and American descent. Eventually there were separate areas for fruits, vegetables, meats, poultry, fish, and wild game. Throughout the nineteenth century and well into the twentieth, the French Market was a cultural, social, and economic crossroads notable both for the variety of produce sold there as well as for the diversity of its venders and customers. For a lengthy description of the French Market from the 1820s, see architect Benjamin Henry Latrobe's discussion in *Impressions Respecting New Orleans: Diaries & Sketches 1818–1820,* ed. Samuel Wilson Jr. (New York: Columbia University Press, 1951), 22. For discussion of municipal markets in general, see Lake Douglas, *Public Spaces, Private Gardens: A History of Designed Landscapes in New Orleans* (Baton Rouge: Louisiana State University Press, 2011), 159–63.

the up-country, asked me, "if Kentucky could beat that?" On telling him it *might* be surpassed, though not far, he remarked that it was Kentucky fed, and brought down by steamboat. His mutton was from Mississippi. The Attakappas and Opelousas beef was small, and though fat, they were not *well fatted*. They looked skinny and flanky, the red, juiceless looking lean, being *overlaid* with heavy coats of tallow. This, it appears to me, is occasioned in part, by irregular supplies of food, but more from the wretched system of in-and-in breeding practiced. The original stock of cattle is good; and even now, some splendid specimens are to be seen. You recollect, I presume, the lot of cattle we saw in a woods pasture in the very richest part of Bourbon county Ky., last summer, and of which you make some sketches.[73] They had run in that pasture for forty years, allowed to breed as they saw fit; and from being originally a good stock of cattle, were when we saw them reduced to queer looking, misshapen creatures, about the size of a full grown billy-goat. Living proofs, that the most abundant supply of food is, of itself, insufficient. This is too much the case with the cattle of the south, so far as I have seen them. The pork, beef and mutton that I saw in market which had been brought down in flat boats looked anything but tempting, and the poultry, oh! All sorts of stock and produce were ruinously low; and I heard many an upcountry trader, curse his luck that had brought him there. Times too, look still more gloomy and squally in prospect, than they even do now.

I had made my calculations for a trip down the LaFourche as far as Thibadouseville [the town of Thibodaux], to spend a few days with my friend Mr. Dinsmore, but my illness here entirely prevents it.

I find that my bantling, the *Western Farmer and Gardener's Almanac* for the present year, has given universal satisfaction even here; and a southern edition for next year so much called for that the attempt must be made.[74]

73. A "wood pasture" is an ancient European land management system, commonly used in Scotland, in which woodland provided shelter and food for grazing cattle together with woodland products, such as timber, for other uses.

74. As editor of the *Western Farmer and Gardener,* Affleck produced an almanac for 1841 (his "bantling" or "young child"), and it was well received in the Ohio River Valley region; no copies are known to exist. This experience certainly foreshadowed and inspired his interest in producing an almanac for the Natchez/New Orleans region. Having moved to the Natchez area and established business ties in New Orleans, Affleck first edited an almanac already established in New Orleans (*Norman's Southern Agricultural Almanac for 1847*), then started one of his own,

You may judge of how much I have felt encouraged and flattered by this.

Yours, &c.

--------◆--------

### To Charles Foster, Esq: A Stroll in the Woods near Baton Rouge[75]
#### *Friday Evening, 28th January, 1842*

A very few hours sufficed to tire me of the dull monotony of this beautifully situated, old fashioned, frenchified, still little place; so picking up my gun, I strolled into the woods.

What a delightful change! Thanks to the enterprising character of the inhabitants of Baton Rouge, I found the woods for miles around almost entirely in that condition I love best—in a state of nature. Not a twig cut; every tree and bush festooned with creepers; numerous open glades, beautifully sodded with Bermuda grass, scattered about with an occasional spreading Magnolia, or oak or that beautiful half evergreen, half pendulous character, so common in the South, standing in the centre; the logs covered with rich mosses new to me; and the whole alive with birds and insects, many of which I had never seen before.

Wandering leisurely on I came to a lonely spot, on the high bank of a little rivulet, where a moss-covered log offered a most tempting lounge. Here I sat me down; threw off my coat, and neckcloth, for the day was warm; laid aside my gun, and cast my eyes around me.

What an endless variety was here! My couch alone would afford a naturalist a month's study. I, though but an indifferent botanist, discovered some five or six varieties of mosses, each of which was in itself a study. Insects innumerable were scattered about, of whose nature and habits I knew almost nothing. On turning over a sheet of the mossy covering, I found four varieties of shells, old acquaintances, and hailing them as such, I carefully replaced their shelter. Some movement I made roused a lizard from his lair, and with the speed of light, he dashed along the log to its further end, where stopping, he afforded me an opportunity of scanning him closely, and re-

---

published in the late 1840s and through the 1850s, in New Orleans. Editions from 1860–61 were issued from Texas.

75. From *Western Farmer and Gardener,* 555–57.

ally he was no beauty! The gaudy and rich colors of many of his tribe, and which he altogether lacked, redeem their loathsome form and render them even objects of beauty. I once kept one a prisoner for many months, whose coat was one of the most resplendent colors, and such as all the skill of the painter could not imitate. The little fellow became so tame as to take flies from my hand, and to recognize my approach—on that of a stranger, he instantly concealed himself. Some accident befell his cage, and I lost him. But this is a digression.

Under my feet was a carpet of violets—pretty blue violets, and of the sweet, modest little wood-sorrel, both in full bloom.

Overhead, my canopy was formed of magnolias and of hollys. The former the most stately and splendid of evergreens—what would I not give to see it covered with its snowy blossoms! The latter now in its fullest beauty; its rich and delicate coat of green, so different from the deep, dark livery of its neighbor, the magnolia, spotted over with numerous bunches of bright scarlet berries. Taller than either, and stretching its gnarled limbs far towards heaven, leafless and grand, stood the swamp-gum—its every twig and branch covered with the long waving festoons of Spanish moss—gloomy and gray, casting an almost sepulchral shadow o'er the tree. I like it—at first I did not—it made me melancholy to look on it—but that soon wore off, and now I like it.

In the rivulet at my feet, sported myriads of the minnow tribe; whilst on its moist banks, crowds of butterflies basked in the sun.

All around me the little birds twittered—those of God's creatures (saving, ever and always, woman) in whose society I most delight. Many of them were strangers, whose acquaintance I would gladly have made; but to slay them I could not then, and I had no other means. I recognized the Pewee, Red Bird (Cardinal Grosbeak)[,] hermit and ferruginous Thrushes, the Robin, Cedar bird, Jay, crested and black-capped Titmice, brown Creeper, golden crested Wren, Carolina Wren, red-headed, golden-winged, downy and orange crowned Woodpeckers, the Dove, Goldfinch, Nuthatch, some of the smaller Sylvias, Sparrows of different kinds, Sowhee, and Bluebird with some others—showing however, but very few varieties that do not also winter with us, and are to be found in shaded nooks in moderate weather.

The Mocking-bird alone, however, gives a novelty and charm to the woods of the South which we can never find where he is lacking. True, they have of late years become more numerous all over the west, and are even found as far north as Detroit. But here, they are in hundreds—here is their home, their favorite abiding place. Naturally bold and courageous, he flits along before you without any seeming fear—flowing with an easy, graceful, careless motion, resembling no living thing so much, as a happy joyous girl waltzing round a room, along, and to the music of her own sweet voice, and from the impulse of her own happy exuberance of spirits—innocent and gay—careless of who admires or who finds fault. Thus it is with the mocker; and as he bears his graceful form along, from tree to tree, his movement and gait differ as much from the flight of others of his tribe, as that of her I have described, from the lengthy stride of the man of business his face turned towards the bank, at five minutes to three!

After several hours of the most perfect enjoyment, I took my way towards town, following the meanderings of the rivulet. Every step brought me food for pleasing thought; and I reached mine inn, a happier, a calmer, and methinks, a better man.

Yours, &c.,

T. A.

# OBSERVATIONS
# FROM MISSISSIPPI

The following lengthy article is an account of the annual agricultural fair convened by the Agricultural, Horticultural, and Botanical Society of Jefferson College, an event that Affleck had previously visited and written about (December 2, 1841). By the time it appeared, Affleck had left Cincinnati and moved to Washington in rural Adams County, married Anna Dunbar Smith, a widow, and moved to her plantation, Ingleside. He took over the management of his wife's properties, which were not in good condition agriculturally or financially. Within a short period of time, Affleck had established Southern Nurseries at Ingleside and had begun to develop business interests in New Orleans and throughout the South.

Affleck gives reports from the committees on stock, sub-committees, cattle, hogs, sheep, and agricultural production with the listings of different contest categories and premiums recommended for the Society's confirmation. Many of the award winners are people Affleck previously mentioned. Categories judged included those for men (livestock) and women (domestic and decorative items). In the men's division, the first premium was a "half eagle" in one category but a "quarter eagle" in others.[76] In the ladies' division, awards included domestic items ("Japan" trays[77]; flower vases; sugar tongs; ornamental plants; one dollar; crop seeds; an annual "bound in Morocco" [leather]; silver pencils; card cases; needle cases; and certificates). He also notes that there were a few farm implements on display but none made locally.

As chairman of the Horticulture Committee, Affleck reports on various awards for wine production, jellies and jams, butter, flowers, needlework, and other items of domestic use. Descriptions of animals exhibited, together with the kitchen and domestic items on display, give insight into farm and domestic life among white residents of rural Adams County in the early 1840s, and they resemble reports elsewhere in this volume of similar agricultural exhibitions and horticultural fairs held in Cincinnati and Philadelphia in 1844. From these reports we gather a better understanding of what Americans of this period were eating, growing, and using in both agricultural production situations and in domestic life. We note, too, how little has changed between agricultural fairs of the 1840s and those of today.

Having now settled in rural Mississippi, Affleck continued to write, and his writings appear in a variety of formats: local newspapers and other agricultural papers; almanacs; and personal and business correspondence. Themes from these early writings continue throughout his later career: advocacy for scientific farming; an interest in increasing agricultural production and efficiency through new management techniques; a concern for advancing the best practices in animal husbandry; his sense of stewardship for the land; and an engaged interest in livestock, bees, horticulture, agriculture, and horticulture. All are informed by national sources and international trends. We also find clues to his personality through his interest in helping others; his sense of the public good; his willingness to observe and make constructive comments; and his appreciation for the work of others. "Friend Affleck," as some correspondents addressed him, was truly a friend to all and an important influence on agricultural practices and horticultural interests in nineteenth-century America.

The writings that follow come from a variety of sources including government publications and local newspapers in Natchez, New Orleans, Houston, and Dallas. Affleck edited existing almanacs (*Western Farmer and Gardener's Almanac for 1842; Norman's Southern Agricultural Almanac for 1847*), then published his own from around 1851 to 1862 (*Affleck's Southern Rural Almanac, and Plantation and Garden Calendar*).

76. The "half eagle" coin was minted between 1795 and 1929. Its value was $5.00 and, after 1837, it contained 0.242 Troy oz. of gold. The "quarter eagle" was worth $2.50 and weighed 0.121 Troy oz.

77. Likely a lacquered tray from Japan.

These publications were supported through selling ads (an on-going concern) that today give first-hand information about domestic goods and supplies, services, and plant availability during that period.

The enduring legacy of these writings is the accounts they give, explicitly through their content and implicitly as descriptions of the general atmosphere of the years leading up to the Civil War and its aftermath. Certainly they show what people were reading in the mid-nineteenth-century Gulf South, and they give descriptions of how things were done. But they also illustrate topics of general concern to the region, how contemporary issues were framed, and what the conversations of the period were about. Taking a broader perspective, we see in these writings an evolutionary process of building community, increasing self-sufficiency, and encouraging economic development through scientific farming, endorsing domestic improvements, stimulating horticultural diversity, promoting technological innovation, and advocating for educational advancement. America moved increasingly in the nineteenth century from a rural economy to an urban one, and with widespread immigration, its population was becoming more culturally diverse. Mid-nineteenth-century issues, often discussed first in agricultural papers, laid the foundation for later attitudes about domestic life, public open spaces, education, and community. And it is from these roots that the profession of landscape architecture emerged in America.

—Ed.

## Proceedings of the Agricultural Fair[78]
### Washington, Miss., Nov. 1, 1843

The Agricultural, Horticultural, and Botanical Society of Jefferson College, met this day pursuant to postponement.

The President stated that Mr. Affleck, by appointment, would deliver an address on "the Nature and Structure of Plants, with reference to Practical Horticulture," which was listened to with a great deal of attention and interest. The society then adjourned to the College grounds for the purpose of viewing the stock, agricultural productions, and other articles offered for exhibition.

On the morning of the 2d, the society met to receive the reports of the different committees. . . .

78. From *Western Farmer and Gardener,* 751–53.

[The Committee on Agricultural Production noted its regret] that in their department they met with but few implements for agricultural purposes, and none of home manufacture.

The committee examined some "comforts,"[79] presented as substitutes for negro blankets. The one which in their judgment best deserves credit for the enterprise and ingenuity of the manufacturer, is that offered by Mrs. Thomas Hall. It is composed entirely of articles grown on the "Cole's Creek" plantation, and is admirably adapted to the purpose for which it was intended. To Mrs. Hall the committee, with pleasure, award the premium—Quarter eagle.

HORTICULTURAL—The Horticultural committee [*Affleck was its chairman —Ed.*] beg leave to report upon the articles included in their department, as follows:

*Best specimen of domestic Wine*—Quarter eagle—To Isaac Dunbar, Esq. This was the only specimen offered; yet, though there was no competition, the first premium was unanimously awarded. This wine is the vintage of 1842; the body and flavor are fine, but its great acidity is objectionable. This, however, Mr. Dunbar's perseverance and enterprise, your committee have no doubt, will ultimately overcome.

---

*[Then follow different categories in this division with the premiums bestowed, including awards for the "Best and most extensive variety of dried and preserved fruits, jellies, and pickles, raised and preserved by the lady presenting them," a "Japan Tray" to Mrs. B. L. C. Wailes. There were "castors, filled entirely with articles of home growth and manufacture, including mustard, cayenne pepper, catsup, vinegar, etc., . . . a pleasing evidence of good management." The prize of a "Japan Tray" "to the lady who shall exhibit the greatest variety of vegetables from her own garden . . ." went to "Mrs. Wm. J. Ferguson whose extraordinary exhibition of vegetables attracted the attention of all." An award was given for the "largest and finest collection of flowers contributed to the decorations," and that collection included "a selection of ten varieties of ornamental plants." Awards were given for "Best Bouquet," "Best Turnips" (also to Mrs. Wm J. Ferguson), "Best six heads of Cabbage," and "Best collection of Dahlias."*

*Next is the "List of premiums awarded by the Committee on Needle Work," of which the new Mrs. Thomas Affleck was the chair. Premiums were awarded for the follow-*

79. Comforters or quilt-like bed covers.

*ing: "largest quantity of needlework"; knit lace; baby linen; embroidery; aprons; lady's mitts and baby socks; patchwork quilts ("No lady looked upon these labored and beautiful quilts, without wishing to be their possessor."); and a hemstitched handkerchief ("a most perfect gossamer"). —Ed.]*

# 4

## From *American Agriculturist* and *Southern Agriculturist*

One of the leading agricultural papers of the mid-nineteenth century was the *American Agriculturist,* started in 1842 in New York City under the editorial direction of brothers A. B. and R. L. Allen. From the outset, the editors envisioned this publication as national in scope, offering to present information to planters, farmers, stock-breeders, and horticulturists across the nation. The Allens, together with another brother, Lewis, were in the livestock business, and in the early 1840s they imported English livestock in an effort to start purebred herds in America. This initiative, together with another advocating for more efficient farm implements, became subjects of discussion in their newspaper.

These two topics were certainly topics of interest to Affleck. His association with E. J. Hooper's *Western Farmer and Gardener* in the early 1840s (including editing its 1842 *Almanac*), the experience in traveling to Mississippi and Louisiana to report on agricultural and livestock matters for the *Western Farmer,* and his recent move from Cincinnati to new markets in rural Mississippi made Affleck well qualified for involvement with the Allen brothers and their *American Agriculturist.*

Other important American agricultural figures associated with the *American Agriculturist* from the early 1850s were reformer Solon Robinson (1803–80) and writer-publisher Orange Judd (1822–80), arguably two of the most important American agricultural figures of the nineteenth century's second half. Robinson contributed to the *Cultivator* and became agricultural editor for the *New York Tribune,* and Judd had studied agricultural chemistry at Yale with Professor John Pitkin Norton (1822–52), an early advocate of scientific farming.[1] In 1856, Judd became sole owner and editor of the

---

1. Judd, Norton, and Frederick Law Olmsted were contemporaries. According to Laura Wood Roper's authoritative biography of Olmsted, Norton was an "old acquaintance" of Olmsted's and gave Olmsted letters of introduction to Scottish and Irish farmers useful in his walking tour

*American Agriculturist.* At the time, it was one of the most popular and influential agricultural journals of its day, a position it would retain for years as it expanded and absorbed smaller periodicals.[2]

From its beginning, the *American Agriculturist* was unique for many reasons. Among them were its efforts to appeal to a national audience; an emphasis on European agricultural advances as models for its American subscribers; abundant and excellent illustrations; and the inclusion of articles of interest to adolescent boys and girls—a feature soon copied by other magazines. Under Judd's leadership, editorial standards remained high, with all advertisements scrutinized for honesty and accuracy prior to acceptance. Obviously, what the Allens started and Judd continued was something the public wanted: three years after its founding, the journal's circulation was around 10,000; by 1859, under Judd's leadership, it had over 45,000 subscribers.[3]

Affleck's contributions to the *American Agriculturist* given here are from 1843–44: one article appeared in the *American Agriculturalist's Almanac for 1844* (New York, 1843) and two articles appeared in 1844 (August and November). The latter two were later collected with three other articles as the journal's Volume 3 (1845). As in articles that appeared earlier in the *Western Farmer,* the subjects are diverse: instructions for growing staple crops and fruits; queries about remedies for sick dogs and cats; and descriptions of "Southern Agricultural Implements." All, save the one on "Dogs and Cats," relate to issues about agricultural conditions of the South, relevant to other farmers in the South and edifying, perhaps, to other subscribers as well.

Given first is Affleck's description of the plantation garden, written in 1843 after Affleck was established in Washington, Mississippi. Worth noting is that this essay is about the utilitarian vegetable garden and not the ornamental flower garden. In fact, ornamental gardens connected with plantation houses were the exception rather than the rule.[4]

---

of England in 1850 observing agricultural practices. See Roper, *FLO,* 65. See also Charles C. McLaughlin and Charles E. Beveridge, eds., *The Papers of Frederick Law Olmsted,* Vol. I, *The Formative Years* (Baltimore: Johns Hopkins University Press, 1977), 216, 237, 341, 342.

2. In its first thirty years, according to Mott, *A History of American Magazines, 1741–1850,* 731, the *American Agriculturist* absorbed twenty-six other periodicals.

3. For general discussion, see Demaree, *The American Agricultural Press,* 348–51.

4. See James R. Cothran, *Gardens and Historic Plants of the Antebellum South* (Greenville: University of South Carolina Press, 2003), 112–15.

An account of Natchez from 1836, written "By a Yankee" (Joseph Holt Ingraham), notes that "[V]ery few of the planters' villas, even within a few miles of Natchez, are adorned with surrounding ornamental shrubbery walks, or any other artificial auxiliaries to the natural scenery, except a few shade trees and a narrow, graveled avenue from the gate to the house." Even though, as Ingraham continues, the planter "may inhabit a building that would grace an English park, the grounds and scenery about it, with the exception of a paling enclosing a green yard, are suffered to remain in their pristine rudeness." There are at least three reasons for this. One, according to Ingraham, is that "planters are not a showy and stylish class, but a plain, practical body of men, who, in general, regard comfort, and conformity to old habits, rather than display and fashionable innovations." He also notes that "one of the finest private gardens in the United States . . . is in the south, and within two hours ride of Natchez," but, he continues, "as a general rule, southerners, with the exception of the cultivation of a few plants in a front yard, pay little regard to horticulture," particularly as a science, but "they are passionately fond of flowers."[5] A second reason, particularly applicable to the Natchez region, is that wealthy landowners usually had several plantation properties in agricultural production in rural areas, but their main residences were in town on smaller parcels of land. Overseers lived on the plantations, usually in modest structures with little ornamentation. Finally, farmers in the South often left homesteads when the soil became depleted and moved west; therefore there was little incentive to plant ornamental gardens since there was the possibility of leaving in the future. In the absence of such established gardens, settlers would elect instead to plant showy annuals and plants that grew quickly for more immediate effects.

If ornamental gardens were more the exception than the rule, one garden all settlers planted, regardless of economic situation or location, was the vegetable garden, and Affleck's discussion of this garden type had immediate application to the needs of this region's residents. He discusses here the cultivation of vegetables commonly grown in the rural South and, by extension, gives a description of the cuisine of the mid-nineteenth-century region's table.

5. Joseph Holt Ingraham, *The South-West: By a Yankee*, Vol. II (Ann Arbor: University Microfilms, Inc., 1966), 100, 102, 114–15. Ingraham does not identify the private garden described as being "one of the finest . . . in the United States."

Plantations, as agricultural enterprises, were self-sufficient in producing food for those who lived and worked there, and obviously it was in the plantation owner's (and overseer's) best interest to keep all—both family and field hands—in good health. Smaller plantation vegetable gardens might be related to the "big house" while larger vegetable gardens would be located near out-buildings (kitchen, smokehouse, stable, barn, workers' cabins, etc.). Both grew vegetables and herbs for culinary and medicinal use by the plantation's residents.

The vegetable garden Affleck describes here is "a primary object on every plantation." Having such a garden for the "laborers," he notes, adds to their "health and comfort" because "[W]holesome, well-cooked vegetables are preferable to anything else, during hot weather." And, he continues, this garden should "be proportioned in extent to the number to be supplied from it." Directions are given for its spatial organization as well as for its preparation and cultivation. Then, Affleck gives descriptions of eleven common vegetables, staples of regional cuisine. Obviously these vegetables are not the only vegetables that were grown; many other plants would have been commonly grown on the plantation—for instance, corn for cattle, sorghum, sugarcane, and rice perhaps—but this brief essay gives a basic description of the contents of a mid-nineteenth-century plantation vegetable garden.

Given next is the first of a two-part discussion of "Southern Agricultural Implements" (August and October, 1844). Such discussions of farm implements were common, and often editors and contributing writers wrote glowing (and dubious) reports on new farm equipment and supplies in exchange for payment by the manufacturer, a practice known as "puffing." Some editors, including the *American Agriculturist*'s A. B. Allen, insisted upon testing all products before endorsing them. Affleck offered, on some occasions, to write about products if manufacturers would send him the product for testing, and sometimes Affleck acted as an agent for manufacturers and their products. For someone of Affleck's influence and wide circle of contacts, this would certainly make good business sense, given the relatively undeveloped reach of advertising in the mid-nineteenth century. Reading this article, we do not get the sense that Affleck is "puffing" or being unnecessarily effusive in his assessments. Rather, we sense that his opinions are rooted in experience and objectivity, and that he is instead seeking to demonstrate to northern suppliers the potential for new southern markets should they create tools designed for southern growing conditions (soils, weather) and crops

(notably cotton). The second installment, not reproduced here, goes into detail on the cultivation of cotton, obviously drawn from Affleck's personal experience as a cotton planter, and he continues to suggest that manufacturers would do well to send samples of their tools for display and testing in southern markets, promising in return to send "a full report of the trial."

The next article from the *American Agriculturist* shows a different side of Affleck, one that is concerned with the health of his dogs and cats, which were always part of a plantation's workings. He noted in the previous article on agricultural implements that "everything on a *cotton plantation,* capable of work, *must work,*" and that would include dogs and cats, even if they were merely pets. An inexplicable ailment is proving fatal to his cats, and he seeks help from others in knowing what the problem is and how to treat it. Here we find Affleck asking (again) for the expertise of others, in much the same way that he freely offers advice on how to deal with problematic situations. His remedy seems almost as gruesome as the ailment itself; nevertheless, his pain is evident even if it is expressed in a morbidly ironic way.

For brief discussion of the *Southern Agriculturalist,* see introduction to the cotton moth a few pages hence.

—Ed.

## The Plantation Garden for the South
### (from the *American Agriculturist's Almanac for 1844*)

### By T. Affleck.

These directions for the management of the Plantation Garden, being prepared for the latitude of Natchez, Miss., can very easily be adapted to a degree or two farther north or south.[6] The garden is a primary object on every plantation. Much is saved by it; and much added to the health and comfort of the laborers. Wholesome, well-cooked vegetables are preferable to anything else, during hot weather. The garden must be proportioned in extent to the number to be supplied from it. For one of some size, instead of a spot laid off in small beds, to be cultivated exclusively with the spade and hoe, select a piece of good ground, no matter what the exposure. Shape, if possible, an oblong square; run one main center walk or road lengthwise; add such others as may be thought requisite; and enclose the whole with a good and sufficient fence. Even though naturally very rich, add a coat of well-rotted manure, as early as practicable in the winter; and immediately turn it under, by running two good plows in the furrow, one behind the other—thus plowing it to the depth of ten inches, or as deep as the soil will admit of, even turning up a little of the subsoil, if not positively bad. When in this rough state even a slight freezing is of great advantage. As the ground is needed for planting, give a top-dressing of manure or rich compost, turning it under with a light plow; and if at all cloddy, run the harrow over it.

As more correct and particular directions can be given, and with less repetition, where each variety of vegetables is treated of separately, that plan is here adopted, in preference to giving a monthly calendar.[7]

6. In editions of his *Southern Rural Almanac* of the 1850s, Affleck gives separate horticultural instructions for Natchez and New Orleans; the 1860 *Almanac* gives instructions for "Natchez, Central Texas &c." regions.

7. Included elsewhere in this almanac, however, are "Monthly Calendars" for the "Northern Garden" and the "Southern Garden." It is noted (38) that since a "great portion of the directions given in the Northern Calendar, for each month in the year, will apply to the South, it is not deemed necessary to recapitulate them. . . . The chief differences consist in the seasons in which these operations are performed, and the cultivation of cotton, rice, sugar-cane, hemp and tobacco. . . . For many of the hints in this calendar, we are indebted to the communications of that truly practical farmer and planter, Dr. M. W. Phillips of Mississippi, which have appeared from time to time in the *American Agriculturist*."

POTATOES. The sweet and the Irish potato are vegetables of great importance. They are cultivated here in the same manner as in the middle states,
and should be planted as early as March or April.

TURNIPS are sown from the 20th of July to last of September—sowing
three or four separate patches, at as many different times. They are usually
sown broad-cast, but would pay well for the trouble of drilling and tending.
The turnip patch is most commonly enriched by penning the cows on the
spot intended for it—but a piece of newly cleared ground is better, producing sweeter roots and fewer weeds.

CABBAGES are produced abundantly in the south, if properly managed,
and are the favorite vegetable on a plantation. They head best on old land
enriched with stable manure. For early spring use, make several sowings of
seed from the middle of August to first of October, of early York, sugar loaf,
&c. During very cold weather, protect the young plants with pine-boughs,
or magnolia leaves; or with stiff brush laid between the rows and covered
with corn stalks and other litter. Plant out early in February. For summer
use, sow in January, protecting as above; plant out when large enough. For
winter use, sow drumhead or other large sorts, in April; let them stand in
the nursery beds all summer, when they will run up a tall stem; during the
rains in August, set them out in rich ground, laying their long stems in so
deep as just to leave their heads out of the ground. If planted out sooner they
will rot; if sowed much later than April, they will not head; and the roots
being placed tolerably deep in the ground, enables the plants to stand the
Autumnal drought. Plants from seeds grown in the south will not head.

OKRA.[8] A large mess of okra soup (called gumbo) should be served on every plantation at least four days in the week, while the vegetable is in season.
The pods are gathered while still tender enough to be cut by the thumb nail:
cut into thin slices, and with tomatoes, pepper &c. added to the rations of
meat, forms a rich, mucilaginous soup. It is planted about the first of March,
in drills four feet apart, leaving a plant every two and a half or three feet, if
the ground is rich, which it should be.

PEAS. Although the dwarf, marrowfat, charlton &c. are occasionally
grown in sufficient quantity for plantation use, it is but rarely.[9] They would

8. Okra is commonly thought to have been introduced into North America by natives of West
Africa. The name *okra* is derived from a word in a Nigerian language.

9. Marrowfat peas are green peas that have been allowed to dry in the field, rather than being picked and eaten when young and tender. It is unclear what variety the "charlton" pea is. It is

form an excellent and wholesome addition to the rations. The crowder and common cow peas being of easy culture, requiring no sticks, being great bearers, and lasting all summer, are indispensable. In winter, the ripe peas form a fine variety. They are planted at any time from the 1st of February to the last of July, either among the corn or alone, in drills 3 feet apart, leaving a plant every foot.

BEANS. Kidney or snap beans are planted in succession during March, April and May—either in hills two and a half feet apart or in rows three feet apart, leaving a plant every four inches. The little white bunch bean sent from the north in such quantities can be raised in the south as easily as any other sort.

LIMA BEANS, or butter beans are grown in hills four feet apart, first planting a stout pole in the hill; plant first of April; leave three to four plants; or they are drilled along the walks, first forming a rough arbor of stakes or of canes for them to run on. They are easily cultivated—procuring and planting the stakes being the principal labor—and are very productive and nutritious.

TOMATOES are indispensable. Sow the seed, in a bed that can be protected, early in February. Plant out, as soon as there is no longer danger of frost, in rows four feet apart, a plant every two and a half feet. A few seeds must be sown about last of April, and again about last of May, to bear until frost; the early plantings will cease to bear by August.

ONIONS AND SCALLIONS [green onions] ought to be cultivated in considerable quantity. They are of easy culture and favorites with the people. Bunches of scallions may be divided and set out in rows at any time from September to March. Onion seed is sown in drills during the fall or early winter, and are drawed [pulled] while young and used as scallions—leaving enough of plants to occupy the ground, where they will bulb.

SQUASH. Of this there are two sorts, with many sub-varieties—the summer bush, and the running squash. The former will produce the greater number on the smallest spaces of ground—the latter, however, continue longer in bearing. Plant toward the end of March, and again about the middle of April the bush sorts in hills three feet apart, leaving one plant in a hill;

---

advertised in 1765 in the Boston *Evening Post* newspaper, and it appears often in late-nineteenth-century horticultural journals. For instance, the *Journal of Horticulture, Cottage Gardener, and Country Gentlemen* 23 (London, 1872): 380, an English horticultural journal, notes that it is also known as "hotspur" or "hots."

the running squash in hills seven feet apart, leaving two plants. The Kentucky cushaw, a large striped, crooked-neck sort can be kept, with a slight protection, all winter. A good supply of squash is desirable, as a wholesome and favorite vegetable; it will moreover prevent your people using young, green pumpkins, which are very unwholesome. As the squash become fit for use, they must be picked off, or the plants will soon cease to bear.

MUSTARD, which may be sown broadcast, and tolerably thin, the seed being very small, in October or November, on a piece of good ground. Mustard makes a wholesome and favorite dish all winter, and early in the spring, boiled with a piece of pickled pork. Like turnips, when sown for the same purpose, it requires no cultivation, if the ground is tolerably clean.[10]

————————◆————————

## Southern Agricultural Implements
### (August 1844)

The consideration you have shown for the wants of the south, in your frequent articles upon implements suited to us, induces me to add my mite of information. I sent you a number of the Concordia Intelligencer, containing the reports of committees at our last agricultural show, and since then wrote you at some length on the trials of implements. Let me again urge you to impress upon your manufacturers of implements and machinery, and especially of plows, the great advantage they would derive, and the vast market they would open to themselves, by forwarding to your shows specimens of the articles they make. It is the determination of some few planters of us, here, to agitate the subject until we are supplied with such as we ought to have, and thus be enabled to meet the present low prices of cotton by an economy of labor.

Our heaviest item of plantation expense, is that for wrought-iron work; particularly to those, who like myself, have no blacksmith of their own. Until very recently, the most simple kind of work in iron cost 25 cents per pound—now it costs a general average of 18 cents. Even this I consider enormous. I should be glad to find a substitute as far as possible, in cast-iron. An ex-

10. Many vegetables were grown for their leaves (or "greens") such as mustard, turnip, collard, kale, carrots, and beets, all of which were cooked as described here.

cellent foundry recently established in Natchez, and which, *as yet,* charges only from 4 to 4½ cents per pound for castings, has enabled me to carry out my plans. And here I want information and advice. I infer from the fact of a people as sagacious and saving as the farmers of New England using cast-iron implements almost explosively, that they must answer every purpose and be more economical. In Scotland, too, they are used to a great extent. But I have here such reiterated assurances that they will not answer, that I almost at times doubt about my own judgment. The objections are, the liability to break, extra weight, want of sharpness, and impossibility of supplying the way, and so on. Now tell us, how is all this?

We require, or rather we should have, a variety of implements. The great cost of everything but the common plow and hand-hoe has prevented aught else being used. In some neighborhoods where there may happen to be good and ingenious blacksmiths, other implements are occasionally employed—such as the bull-tongue, or narrow single-shovel plow, for running on each side of the corn-row at the first tending; triangular, one-horse harrows; scufflers, or rough cultivators, with three to five teeth, somewhat like a common hoe; sweeps, skimmers, or spread-eagle or buzzard plows; double-shovels, and double half-shovels, &c., &c. the cost and the difficulty of getting a good article, *sure to run well,* which is a great difficulty with wrought-iron implements not made by a master hand, has prevented even these improvements being commonly used. Cast-iron articles, with stocks so simple as to be easily made on the plantation, will I think meet those objections. Such a thing as Wilkie's horse-hoe will not suit us at all—too lengthy, complex, and expensive to put in the hands of a negro. Mr. Thorpe's three-share plow has the same objections, with the additional one of too great weight. Our teams cannot drag along such a load of wood and iron as can your stronger animals, in a cooler clime—*one animal only* can be used to do the tending of the crop, which is done during the hottest season of the year, when one of our average-sized mules (which form our best and most economical teams) drags an implement weighing 50 or 60 pounds, ten hours in the day, between rows of tall cotton, corn, or cane, he has as much as he can possible stand, and more, in many cases. Yet *he must do it*—there can be no cessation of work—everything on a *cotton plantation,* capable of working, *must work.*

Your northern-made implements cost us too much by the time we get them, passing through so many hands; and if we order them direct, we are buying a "pig in a poke," where we have only a published account and description to go by. It is on this account we are so desirous of seeing your northern implements well represented at our trial in the fall. *You* require a *heavy* as well as a strong plow, with great length of share and landside to make it run steady in stony land. We *require* nothing of the kind. A plow, to suit us, must have *size,* and yet be light for man and beast—easy to handle among stumps and roots, on steep and short hill-sides, and among the young, delicate, and easily-injured cotton plants. When you send us a *light* plow, they are so small and slight as to be almost worthless—nothing but the merest *pony* will suit to hitch to them. We have no stones to trouble us, rarely any sod to cut—nothing but weeds and trash on light mellow earth— unless where almost ruined by being trodden by stock. Hence we require a plow that throws dirt well, not easily chokes, and which turns a furrow 10 inches wide, and 5 to 6 inches deep, with two average-sized mules.

The best plow I have met with, *for all work,* is "Hall's Improved Peacock," № 2, made in Pittsburg. It is a good sized breaking-plow, for two common-sized mules; covers up trash well, and of course, ridges well; cost I think, $6.50 or $7; is strong, yet light and handy. For a regular breaking-plow, on land not *too* hilly, and with a moderately strong team, and particularly where there is a stiff sod, or if the ground has been trodden by stock, I have seen nothing to equal the Eagle Plow of Ruggles, Nourse & Mason, with coulter and wheel—both of which are indispensable—which was tried at our last show.[11] I afterward purchased it at $13 (too high a price) and find it does excellent work. I sincerely hope that this and other firms will see fit to forward for trial here in October, specimens of their different sized plows and other implements; the more as you will observe that the Messrs. Holds (of Natchez and Boston) will convey them, for that purpose, from Boston free of cost. This would open up a new and extensive market to them. Corn and cob-crackers, fanning-mills, grist-mills, thrashing-machines, straw and stalk-cutters, corn-shellers, grain-cradles, steel hoes, gin-stands, &c., are all in demand. If the makers of Batchelder's planting-machine have improved that excellent implement, so as to give it a little more strength, and to permit the attendant to see the corn as it drops, he may send one to our trial with the certainty of introducing it here. If it operates as the specimen one I had tried at Cincinnati, I will agree to purchase the one sent, and hand over the price

11. Ruggles, Nourse & Mason was a wholesale/retail dealer in seeds, plants, trees, agricultural implements, and similar items in Boston, at Quincy Hall over Quincy Market.

to the president of our society to be remitted to him. The objection made to it in the west, that it will not drop in hills that can be tended both ways, is no objection here, as we tend everything in drills. I know of no implement that would be of equal value to the planter.

THOMAS AFFLECK
*Ingleside, Adams Co., Miss.,*
*27th May, 1844*

---◆---

## Dogs and Cats
## (November 1844)

I suffer great loss in two kinds of stock here, which, if not of as great value as Durham cattle, are yet even more indispensably useful—I mean in *dogs and cats!* During the two years and a half that I have resided here, we have not been able to keep a single cat; they have all died in convulsions, and all in the same singular manner. They are attacked with violent shivering, seem in great agony, mewing and struggling, each fit becoming more and more violent, until they die. Can you or any of your correspondents give us the cause and remedy.* I have been told that *bleeding,* by cutting off a piece of the tail each time they have a fit, will ultimately cure them. I am now trying it, *economizing the tail* as much as possible, that it may get a fair trial. So far the kittens have recovered, when thus treated. Some other cure would be preferable, as a *bobtailed* cat is rather an unsightly object. Still, better, even a bobtailed one than none.

I have also lost several valuable dogs within the past year, in somewhat the same way. Two of those I lost were *very valuable* terriers. They begin by going about as if in pain, and evidently not thriving. In a few days they commence with a sharp, keen, constantly reiterated bark, which they keep up, day and night, concealing themselves in some dark corner for a week or so, when they die. Others have dropped down, when apparently in good health, in a violent convulsive fit, having one fit after another, in rapid succession, until they die. I meet with no loss, of this kind, that grieves and annoys me so much as that of a favorite dog, and would be glad to hear of some cure or preventative.

I would say to your correspondent, S.S., that we have had our trees and shrubs almost ruined this year by myriads of a dark-brown aphid, yet we have no sight of the yellows.

THOMAS AFFLECK

*Washington, Miss., Oct, 1844*

*We cannot tell the cause of this disease, unless it be an overeating of rats or animal food, but the remedy we have generally found successful, was, to administer pretty strong doses of warm catmint tea. As a preventative, we supply our cats with all the milk they will drink and what vegetable food they will eat, much as bread, potatoes, &c. We also occasionally give them a dish of fresh fish, well cooked, of which they are extremely fond. All animals should have a variety of food when possible to obtain it.

## ON THE COTTON MOTH

In addition to information about agricultural matters, Affleck also wrote about insect pests such as ants, moths, and termites, that plagued farmers in the Southwest and corresponded with those, as has been mentioned, who could instruct him in the management of such pests. Worth noting here is that Affleck's son, Isaac Dunbar Affleck (1844–1919), became an entomologist and studied ants; part of his findings on the Texas agricultural ant and the cutting ant are contained in H. M. McCook's *The Natural History of the Agricultural Ant of Texas* (1879).[12]

The following article appeared in the October 1846 edition of *Southern Agriculturist* (Vol. 6, 387–89) but apparently was first published in the New Orleans *Commercial Times*. The *Southern Agriculturist,* from Charleston, South Carolina, existed between 1828 and 1846 and was one of the few agricultural papers published in the South. It was aimed at the small planter and, unlike other papers, its editor did not advocate an agenda, express his opinions, or write much for the paper. It never achieved wide

12. McCook notes (15) that "Mr. Affleck adds to a well-educated mind practical skills in agriculture." See also Samuel Wood Geiser, "Notes on Some Workers in Texas Entomology, 1839–1880," *Southwestern Historical Quarterly* 49 (April 1946): 593-98. Geiser, author of *Horticulture and Horticulturists in Early Texas* (Dallas: Southern Methodist University, 1945), in which a biography of Thomas Affleck is given, lists in this article both father Thomas and son Isaac as being early entomologists in Texas.

success or circulation, and never attracted much interest, in the form of contributions, from its readers; hence much of its content came, as in the articles here, from
other sources.[13] It demonstrates the descriptive, straightforward approach that Affleck took in describing situations that he felt farmers should be made aware of, and
it closes with the suggestion that farmers should cooperate with each other, meet on
a regular basis to discuss matters of common concern (such as how to deal with this
insect pest), record their observations, and then forward them on for dissemination—
by Affleck—to a larger audience.

Immediately following this article is a much longer one on the same subject, "The
Cotton Moth, or Caterpillar—Its Effects upon the Crop." There is no indication that it
had appeared earlier in the *Commercial Times,* but there is, however, a note naming
Affleck "our agricultural correspondent." At the writing of the first article (August
1846), Affleck feared that the pest would destroy the current crop, and by the time of
the second article (one or two months later), that fear had been realized. He notes
that "*At this moment those immense fields are one great waste.* Not a leaf to be seen.
Everything upon which the mandible of the work could operate being consumed, they
are now travelling in search of more food. And thus it will be with every crop in the
country, so far as I can learn."

Affleck's second article discussed the moth "in more detail," providing information he had obtained from other sources about its history in other regions, including
the Bahamas and West Indies, Africa, South America, and other states in the South.
He notes that farmers in Maryland and Virginia have used domestic fowl, particularly turkeys, to control the pests. "Turkeys are observed to have a remarkable appetite for the larvæ of the cotton moth, and devour prodigious quantities of them." He
also notes that "without a more general and united effort on the part of the planting
community, than we can possibly expect to see, little can be hoped for" in terms of
eradicating the pest. Affleck then gives five remedies successfully used thus far by
John Townsend, of South Carolina:

1. His people searched for and killed both the worm and the chrysalis of the
   first brood.
2. On the appearance of the second brood, he scattered corn on the field to
   invite the notice of the birds, and while they depredated on the worms on
   the tops of the stalks and their upper limbs, the turkeys destroyed the enemy on the lower branches.

13. See Demaree, *The American Agricultural Press,* 356–58.

3. When in the Aurelia state, the negroes crushed them between their fingers.

4. Some patches of cotton, where the caterpillars were very thick, and the birds and turkeys could not get access to them, were destroyed.

5. The tops of the plants, and the ends of all the tender and luxuriant branches, where the eggs of the butterfly are usually deposited, were cut off.

Affleck closes his second article with an account of his own experience:

In 1844, I suggested the use of torches or small bright fires, scattered over the fields, at night, into which the moth would fly and be destroyed. These have been extensively employed this year, to the destruction of myriads of moths. I shall, in the future, lay in a few barrels of rosin[14] as part of the regular plantation supplies, to be added to the fires which shall be kept burning at intervals through the entire summer.

He also suggests here the "judicious rotation of crops would do much to render innocuous the visits of the caterpillar." Worth noting is that Affleck hoped the local government would take action: "The attention of the Legislature of Louisiana—I have little hope of that of Mississippi—should be called to this subject, and that of the boll work; and means used by them to try every method of checking the evil." Affleck closes his article saying that he "would feel particularly obliged by any information on the subject of either this or the caterpillar, addressed to me at Washington (Miss.)."

## The Caterpillar

From *The Southern Agriculturist*
Vol. 6 for the year 1846

From the *New-Orleans Commercial Times.*

### THE CATERPILLAR

We beg to call particular attention to the subjoined communication from our valued agricultural correspondent, Thomas Affleck, Esq., in relation to the army worm, which has recently committed, and is still inflicting severe

14. Rosin is the sap of pine trees, distilled and dried to a brittle solid of many uses. It is extremely flammable, burning with a smoky flame that would certainly draw—and destroy—moths.

injury in the cotton-field. The information which he conveys is extremely discouraging for the planter.

Ingleside, Adams County, Miss.

August 29, 1846

*To the Editors of the Commercial Times:*

Finding much difference of opinion as to the nature and character of the worm now ravaging our cotton fields, the extent of the injury already done, the probabilities of their further increase, and of the crop being ultimately lessened in amount by them; and finding, further, that an opinion expressed by me as to the effect of the worm's ravage has been misunderstood, I shall in this instance speak in the first person, leaving you to add editorially, what further information you may possess from other quarters.

It is now some four weeks since the appearance of *the caterpillar* was spoken of. It was, at that time, supposed to be a mistake—that some other work had been taken from the true cotton devourer. I made every inquiry in my power relative to the worms spoken of, and took an early opportunity of seeing them. They are the same as those we had in such myriads in 1844, and such I pronounced them, holding out no hopes of escaping injury this year, but advising all to *console themselves with the hope* that the stripping of the leaf might aid the cotton in opening this backward, wet season.

At this time, the worm is in every cotton-field in this region of country, and, I have no doubt, in every county in the State. The present *crop* of them is not sufficiently numerous to do material harm—there being not more perhaps than from half a dozen to a dozen on a plant. These have mostly attained the chrysalis and moth state. The latter is to be found, in the greatest numbers, not only in the cotton fields, but in the woods and swamps, and hills—miles from a stalk of cotton. They travel rapidly, and to a great distance. *I have no hope, whatever that the plant will escape total destruction from the myriads of worms that must of necessity appear in a week or two from the eggs laid by these moths!*

To prove that I am not needlessly alarmed, let us look at their Natural History—it must be a mere sketch—I have not leisure, nor am I yet fully prepared to give a full and scientific account of them.

The present insect is the *Noctuaxylina*,[15] a night flying or owlet moth—I

15. Usually given as *Noctua xylina* or cotton moth; discussion included in Affleck's *Almanac, 1851*, 50–51, also in the *Southern Agriculturist* (1828) and subsequent publications on growing cotton.

think belonging more properly to the Mamestradæ—of a beautiful green-ish grey, with bronze shading: on each outer wing there are two small white spots, shaded with bronze, near the shoulder and in a line with the edge; and lower down, a large kidney-shaped black or brown spot, shaded with white. Several wavy lines of purple cross the outer wing, which has also a fringe of the same color in the inner edge, and a fringy grey and purple at the end. The body is thick, and tapers to the end. The female is larger than the male, but they are otherwise much alike. The female deposits her eggs on the leaf (I am not fully satisfied that they are placed *only* there) in clusters; they are round, and whitish or pale green, and quite small. They are hatched in from two to five days, according to the weather, and immediately commence eat-ing the leaves of the plant. They increase rapidly in size, attaining their full growth of one and a half inches in from two to three weeks. They are of a light green color, with longitudinal stripes of yellow on the sides, and along the back two black ones, separated by a very narrow line of white. Some are without the black stripes. They are also studded with small, distinct black spots, from nearly every one of which a black hair grows. They have sixteen legs—one pair behind, eight in the centre of the body, and six pro-legs. They elevate the front half of the body, when at rest, giving it a continued motion from side to side. They give forth, when in numbers in a field, a peculiar sweetish odor, readily recognizable by the observant planter. During the lifetime of the worm, it casts its skin at least four times. When it has attained its full growth, it places itself near one of the corners of the leaf, spins a few threads of silk, attaching them to the leaf in such a way as to draw up the edge, which it makes fast to the surface of the leaf, forming a scroll, within which it undergoes its transformation to a *pupa*. This it does in thirty-six to forty-eight hours. The pupa is black and shining. In this state it remains from one to three weeks. I have found the state of the weather influence the change thus far, generally from seven to twelve days—when the perfect in-sect appears, and proceeds to multiply her species. This each female will do to the extent of from two to six hundred or more.

It is thus seen that we cannot entertain a hope. Those moths which have been flying, in such numbers, around the lights in every house, for the last week, have already deposited their eggs, or are now engaged in doing so. They will quickly hatch out, and absolutely over-run us. I have accounts of them in almost every cotton-growing State. The crop is *at least* three weeks later than usual, I think that the incessant rains we have had, and still have, put it back yet another week. The first crop of bolls is rotting very much—what little cotton has gone to market, you will observe, has been pronounced very middling in quality, and even that open slowly. The after-crop the worms will either destroy, or greatly injure, by eating off the calyx leaf or *square*. In ordinary seasons we have a bale to the hand housed by this date—this year, scarcely a pound. The boll or bore-worm, too, a new enemy in the hills, within the last two years, will destroy ten percent. of the young bolls, after they have attained a size at which they are usually considered safe. Of him I shall speak at another time—as also of the checks and preventatives to be used for both.

I am by no means desirous of seeing a panic got up but really think that every cotton-grower should hold on to his present year's crop, until he can form a pretty close opinion as to the amount of damage likely to ensue from the worm. As a general rule, it is always best to take the chances of the market, selling as quickly as the cotton is sent in. But the present year is an exception. By the 1st or 15th of October, we shall be able to know. Good cotton will be a scarce article, at best, this year, and *must* command a high price. If the grower does not realize that price someone else will.

Let me here urge it upon the reading, observing class of planters, to observe those insects closely, and learn during the present season, all they can of their nature and habits, and submit it at once to writing. If they will communicate direct with me, I shall be glad. Let them also endeavor to form a neighborhood "Planter's Club"—proposing to some ten or a dozen of their most intelligent neighbors to do so, meeting once a month at each other's houses, alternately; spend the forenoon in a ramble over the plantation, the owner pointing out and defending his practice; after dinner, let some subject be discussed, such as the *nature and habits of this caterpillar,* from the close, personal observations of *each member*—such subject having been specified at the preceding meeting. Let some one act as secretary, and note down the results, handing his noted to the *Commercial Times,* or the nearest local paper, for publication. You will see at once, how much good might thus be affected. The planter, too, could thus bring to bear, in favor of his business, the immense power and influence of association.

Yours, &c.,

T. A.

# 5

## Affleck's Contribution to a National Discussion of Agriculture

### Agriculture in the State of Mississippi

The following article, from 1849, was addressed to Commissioner of Patents Thomas Ewbank in response to his request for an account of the state of agriculture in Mississippi, and it appeared in the *Report of the Commissioner of Patents for the Year 1849* (Washington, D.C., 1850). Over 11,000 words long, it is one of the longest and most detailed examples of Affleck's published works. Here we find the breadth of Affleck's interest, the depth of his knowledge, and the characteristics of his writing: informative, engaging, and straightforward. He writes with authority about numerous issues related to agricultural cultivation (of grains, grasses, and root crops; cotton and sugarcane); the state of the land; the condition of local livestock; and, briefly, the costs associated with labor, both enslaved and white. In addition, he discusses fertilizers, fruit orchards, and, briefly, grapes.

This article represents how far American agriculture had advanced by the mid-nineteenth century in content, scale, and regional specificity from late-eighteenth-century interests in local subjects to topics of national discussion, directed by articulate and experienced regional voices. Agricultural knowledge had also advanced, from mainly eighteenth-century English books on gardening with generic contents to specific works written by Americans for both regional and national audiences. Much of the debate about agricultural issues was apolitical and sidestepped larger political issues (slavery) that characterized regional approaches to agricultural production. Nevertheless, the nation's progress in agricultural reform—understanding successful growing techniques and efficient land-management practices, improving crops and practices of animal husbandry, and becoming more efficient with resources—can be attributed to journalists like Affleck and to the agricultural press where their observations appeared and their recommendations were discussed.

As previously mentioned, from the late eighteenth century onward, educated and prosperous landowners, elected officials, and community leaders organized agricultural societies for the dissemination of information about advances in agriculture and science. Modeled on European precedents, these regional clublike organizations convened regular meetings, sponsored lectures and exhibitions of farm produce, exchanged information with similar organizations in Europe and elsewhere in America, and often printed proceedings of meetings, lectures, and scientific papers presented by members and guests. These societies were for the most part concentrated in urban population centers but were concerned primarily with the issues of their outlying rural regions. Members understood that the strength of the new nation was based on its agricultural industry, and one of their interests was in establishing a national governmental agency for agricultural matters. George Washington, toward the end of his presidency, recommended establishing a governmental agency to deal with agricultural issues, but there were no agencies at any level of government concerned with agricultural matters until the mid-1830s, and reliable information about national agricultural production, methods, and practices was not readily accessible until the late 1840s.[1]

To support progress in technological innovation, the nation's patent office was created in 1835 within the Department of State, and it was through patents for farm implements that this agency became involved with agricultural matters. Henry Leavitt Ellsworth (1791–1858), an attorney interested in improving agriculture, became the first Commissioner of Patents. Born into a prominent Connecticut family, Henry was a son of Founding Father and third U.S. Supreme Court Chief Justice Oliver Ellsworth. He graduated from Yale in 1811 and left on an extended trip to the "West" (what is now Ohio) to

1. In Washington's speech on the opening of Congress, December 1796, he said, "It will not be doubted, that, with reference either to individual or national welfare, Agriculture is of primary importance. . . . Institutions for promoting it grow up, supported by the public purse; and to what object can it be dedicated with greater propriety? Among the means which have been employed to this end, none have been attended with greater success than the establishment of Boards, composed of proper characters, charged with collecting and diffusing information, and enabled by premiums, and small pecuniary aid, to encourage and assist a spirit of discovery and improvement. . . . Experience accordingly has shown, that they are very cheap instruments of immense national benefits." *Facsimiles of Letters from His Excellency George Washington . . . to Sir John Sinclair . . . on Agricultural and Other Interesting Topics* (Washington: Franklin Knight, 1844).

deal with family property. Later, Ellsworth was appointed by President Andrew Jackson as one of three Commissioners of Indian Tribes, and in 1832 he returned "West" to Ohio, Kentucky, and Missouri to deal with issues related to Jackson's policy of relocating Native Americans into what became the Oklahoma Territory.

When appointed the Commissioner of Patents in 1835, Ellsworth found the patent office in disarray, but he soon devised an orderly system for issuing and documenting patents. With this process established and a keen interest in scientific innovation, Ellsworth was instrumental in working with inventors such as Samuel Colt (the revolver), Samuel Morse (the telegraph), and many others. Through his exposure to the unsettled lands of the West, he became interested in agriculture and agricultural innovation and the role they played in the nation's future. By 1839, Ellsworth had persuaded Congress to establish the Agricultural Division within the Patent Office and secured an allocation of $1,000 for "the collection of agricultural statistics and other agricultural purposes." He collected seeds and plants from abroad and distributed them through members of Congress for their constituents and to agricultural societies for their membership. These organizations, generally structured either as community or statewide groups, counted prosperous landowners, influential leaders, and elected officials as members and were often loosely affiliated with various agricultural periodicals. And while the agricultural press and agricultural societies all shared a dedicated interest in improving agricultural education and increasing awareness of more effective means of production, they all shied away from political discussions that were inextricably linked to issues of agricultural innovation and advancement.

Ellsworth remained Commissioner of Patents for ten years, having created by 1845 a functional national bureau that dealt with both the issuance of patents and national agricultural concerns. After leaving the Patent Office, Ellsworth moved to Indiana and became involved in the sale and settlement of public land. He returned to Connecticut in 1857, serving as an insurance executive until his death.

Ellsworth's interest in aiding agriculture was evident in his annual agriculture reports to Congress and his advocacy for a public depository to preserve and distribute various new seeds and plants, a clerk to collect agricultural statistics, the preparation of statewide reports about crops in different regions, and the application of "scientific farming" principles to farming.[2] The publication of annual reports came as a result of Congressional action in 1847 entitled "An act in addition to the act to promote the progress of science and the useful arts," amending legislation first passed by Congress in 1790.[3] The basis for this action was Article I, Section 8, of the United States Constitution (1788): "The Congress shall have power . . . To promote the progress of science and useful arts, by securing for limited times to authors and inventors the exclusive right to their respective writings and discoveries," also known as the "Copyright Clause." The Act of 1790, the first attempt to implement this provision, created a system to protect inventors' rights through the granting of government patents. Because America's economy at this time was almost entirely based on agriculture, such "writings and discoveries" covered under this clause were related primarily to the agricultural economy, hence the connection between patents (and the Patent Office) and agricultural production in the mid-nineteenth century. Over time, legislative acts based on this section of the Constitution have been enacted, expanded (patents separated from copyrights), amended, and in some cases ultimately interpreted by the United States Supreme Court.

Ellsworth began publishing annual summaries of his office's activities (including office expenditures, patents issued, and a list of patentees with their place of residence) in the late 1830s, and over time more information was added to what was collected as it became available. Such reports established the format and content of Patent Office Reports, which were continued and expanded by subsequent commissioners. Publication of these annual reports on agricultural activities for Congress and the American people, which continue to the present, was a response to the advocacy of agricultural reformers, many of whom were editors of period agricultural papers such as

2. Often derisively termed "book farming" by many, such practices were not universally adopted by Americans, although they were enthusiastically supported by agricultural reformers and the editors of agricultural papers. In general, "scientific farming" had several inter-related principles, all aimed at increasing agricultural production and efficiency: education (to replace traditional means of farming with facts, based on science); technology (efficient equipment); management (the creation of systematic and organized approaches to production based on long-range agricultural and management strategies rather than continuing to use haphazard approaches based on short-term issues); improved species (of both plants and animals to increase productivity); and land stewardship (to avoid depletion of soils through an understanding of soil sciences, water management, irrigation, and additives such as fertilizers).

3. *Annual Report of the Commissioner of Patents, for the Year 1847* (Washington: Wendell and Van Benthuysen, 1848), 3.

the *Western Farmer and Gardener.* These reformers advocated for a national office for agricultural affairs, dissemination and application of scientific techniques for increasing production yields, and for college-level instruction in farming and farm practices at public institutions. When the Commissioner's "Reports" began, the information presented came from an examination of data from the previous year, together with a "laborious examination and condensing of a great number of agricultural papers, reports &c., throughout the Union, together with such other information as could be obtained by recourse to individuals from every section of the country," according to an article in the *Western Farmer.*[4] Contributions from correspondents from throughout the country bear a strong resemblance in both content and style to articles that appeared in the agricultural papers of the period.

To facilitate the gathering of such agricultural statistics from around the country regarding agricultural production, yields, economic information, and other data, circulars were distributed seeking information from those engaged in agricultural work. We find Affleck discussing this effort in the opening paragraphs of the following article. As published in the *Western Farmer and Gardener,* the 1842 report included two tables with, first, production by state of agriculture products (wheat, barley, oats, rye, buckwheat, corn, potatoes, hay, flax and hemp, tobacco, cotton, rice, silk cocoons, sugar, gallons of wine); and second, production of other crops (wool, hops, wax); populations of livestock (mules, cattle, sheep, swine, poultry); values of various products (dairy, orchard, homemade goods); produce (from market gardens, nurseries, and florists); as well as accounts of the number of men employed and the amount of capital invested. Following these tables is a narrative (Ellsworth's annual report) in which he analyzes these statistics and speculates on their meaning. Notable too are his narratives about specific crops (e.g., wheat, barley, oats, rye, etc.), analyzing production, where they are found, and speculating on how improvements might be made.

By the 1846 edition, the commissioner's format for gathering agricultural information and the method of presentation to Congress was established, and while we do not have a copy of the commissioner's questionnaire soliciting information, we do have evidence that it was highly regarded by the agricultural press. According to an unsigned article in the *Western Farmer*

4. "Ellsworth's Report on Agricultural Statistics," *Western Farmer and Gardener* 3 (Cincinnati, 1842): 158–64.

*and Gardener* (June 1846), the commissioner's efforts to gather this information into a report

> is one of the felicitous *inventions* of our age; and ought itself to be patented.
> It is not a dry detail of models and machines; but the embodiment of the
> *industrial* progress of the United States from year to year. Nothing could
> take its place, and no other publication has any resemblance to it. Nor
> do we suppose that it could be got up and circulated except by the United
> States Government. . . . It seems that everything that is said in every news
> paper, journal, agricultural paper, or county political print . . . [is] gath
> ered, abridged if need be, and classified, so that the work is, in some sense,
> a digest of the best agricultural and horticultural information of the year.[5]

Affleck's article comes from the 1849 report, published in 1850. Modestly
titled "Essay on Grasses for the South," this piece covers a much wider range
of topics than just grasses. It replicates the content established by previous
annual reports but goes into much greater detail, reflecting the interests
of Thomas Ewbank, then Commissioner of Patents (1849–52).[6] Reading between the lines in Affleck's article, we get a sense from some of his responses
relating to the "cost of production" of crops and the "average per acre" mentioned in his discussion of rice, that systematic inquiry was beginning, on
the national scale, to gather data related to the costs (and methods) of production with results as metrics useful in evaluating the content and efficiency of agricultural endeavors.

In 1849, the Patent Office was transferred into the newly created Department of the Interior. National interest continued for a separate office for agricultural affairs, sustained through the advocacy of the agricultural press
and a new generation of agricultural reformers. It was realized in 1862 by
President Lincoln and led by a commissioner. Cabinet-level status for the
Department of Agriculture came in 1889, over fifty years after Commissioner
of Patents Ellsworth began the nation's first office for agricultural affairs
and after years of advocacy by agricultural reformers such as Affleck and
his contemporaries in the agricultural press. In recognition for his lead-

5. *Western Farmer and Gardener* 2, no. 11 (Indianapolis, June 1846): 161–62.

6. For a summary of Ewbanks's tenure as Commissioner of Patents, see William A. Bate,
"Thomas Ewbank: Commissioner of Patents, 1849–1852," *Records of the Columbia Historical Society* 49 (Washington, D.C.: Historical Society of Washington, 1974): 111–24.

ership role, Ellsworth is now known as "the Father of the Department of Agriculture."

By November 1849, when Affleck's article appeared, he was established in Mississippi, and the detail with which Affleck discusses his topic indicates a substantial grasp of the subject, no doubt based on his skills of observation, his intellectual curiosity, and his willingness to engage others from whom he might learn. Toward the close of his discussion on fruit trees, Affleck writes: "As it is my intention to give to the world, in a few months, a familiar treatise embracing the entire gardens and orchards in the South, I shall not extend these remarks much farther." Producing such a "treatise" was a lifelong ambition of Affleck's, yet one he did not live to accomplish.[7] Yet in this article addressed to the Commissioner of Patents, as well in all of Affleck's published works, we see a broad intelligence, an eagerness to share information, a curiosity about how and why things grew (and why not), and a recurring, convincing argument for "scientific farming." These were all items on Affleck's agenda, and they are the ones that would occupy most of his professional life.

—Ed.

7. On the other hand, Affleck may be referring here to his *Southern Rural Almanac*. At the time this article appeared, Affleck had edited *Norman's Southern Agricultural Almanac* (1847) and was likely making plans to publish his own almanac. Nevertheless, Affleck later began work on an ambitious multi-volume "treatise," but only one volume, *Hedging and Hedging Plants, in the Southern States* (1869), appeared, published posthumously.

## Essay on Grasses for the South

*Report of the Commissioner of Patents for the Year 1849,*
Part II, Agriculture[8]
Washington, Miss., 15th November, 1849

Dear Sir:—I have often tried to draw up a series of answers to the queries from the Patent Office; but could state so little that was satisfactory, had so few facts as a basis, that I gave up the attempt. I shall try again, however; and have placed the extra circulars sent to me in the hands of those I thought might possibly fill them up, but fear few will do so.

Cannot the several States be induced to require of the tax assessors of each county to fill up a series of condensed points and queries, to be put to each planter assessed? I know of no other way in which you can procure the statistical facts you require with any degree of correctness.

A second edition has been published of the "Plantation Record and Account Books," which I prepared for publication at the request of a publisher in New Orleans.[9] They have come into very general use, and will in a few years afford a mass of the most valuable information, of the very nature required. You will receive herewith a copy of each number for different sized plantations, with the hope that you will give me the benefit of the experience of your office in suggestions for the improvement of the next edition; that the work may be made the means, if possible, of preserving a still greater amount of that information of so much importance to the country. But to the several questions in the circular before me.

8. Washington, D.C.: Office of the Printers to the Senate, 1850, 152–67.

9. In the late 1840s and the 1850s Affleck developed and revised a highly effective means of accounting for agricultural production expenses and returns with his *Plantation Record and Account Book* for cotton and sugarcane plantations. The books were large-format ledgers organized according to the size of the plantation, based on the number of "hands" working the land. There were three versions for cotton plantations (№ 1 for "40 hands or under"; № 2 for "80 hands or under"; № 3 for "120 hands or under") and two for sugar plantations (№ 1 for "80 hands or under"; № 2 for "120 Hands or under"). These ledgers provided a means of keeping track of inventories of stock and implements, expenses, daily records of events, amounts of cotton picked by workers, physician's visits, births and deaths of slaves, and so on. Since such a system had not been devised before, farmers could not easily associate production expenses and production outcomes. Widely available and sold throughout the country, these ledgers were used primarily in the South. The fifth edition of the *Cotton Plantation Record and Account Book* appeared in 1854; the eighth in 1859. From the information collected, one could easily determine statistics such as those sought by the Commissioner of Patents.

*Wheat.*—There is but little grown in this district. In the adjoining county of Jefferson, good crops have been grown, with occasional discouraging failures from long droughts. There is no doubt but wheat can be grown profitably, but not until the wants of the soil are considered and supplied.

*Oats.*—Singular as it may appear, this, looked upon as altogether a northern grain, succeeds well here. I have imported many varieties from Scotland and France, and have grown them successfully; but have found none of them to compare with the variety known as the *Egyptian* or *winter* oat. It has been in the South many years, perhaps fifty, and is *thoroughly acclimated.* Of the importance of acclimation, more anon. I have repeatedly found this grain to weigh 42 lbs. per bushel. The grain is white, large and plump. It is sown in September and October, and even later; ripe during May, according to the season and soil—with me, during the first and second weeks. Affords excellent winter pasture. By planters and overseers generally, the oat is considered to be very inferior to corn as food for mules and horses, during hard work. The cultivation of the oat does not extend much, arising, it is to be feared, from an indifference to improvement or change, and from the trouble attending the cutting and threshing. This variety, the Egyptian, is invaluable, not only as fodder crop, but for winter pasturage and as an auxiliary in improving the land.

It is advantageously sowed amongst the cotton, after the first or second time it is picked over, or amongst late corn; the sweep or cultivator being used to cover. During the winter the stalks are beat down as usual, not at all interfering with the cutting of the oats. If intended for the improvement of the land, hogs or other stock should be turned in when the grain begins to change color, and when they have eaten it pretty clean, plough the stubble under and sow cow-peas; these to be fed *off,* in turn to be followed by oats again or clover.

*Rye.*—I have grown the *Multicole* and the St. Johns-day rye, or *"Seigle de St. Jean,"* imported from England and France[10]; neither of these were su-

10. This variety of rye was introduced into America from western France, and according to Lynn A. Nelson in *Pharsalia: An Environmental Biography of a Southern Plantation 1780–1880* (Athens: University of Georgia Press, 2007), it caused a "minor sensation" in the agricultural press in England and America because of its high yield and resistance to rust (a common malady). A few bushels of seed were imported by the Commissioner of Patents in 1842 and distributed to prominent landowners. Two years later, the owner of Pharsalia Plantation in Virginia's Piedmont region planted a crop and had incredible success, and soon this variety of rye was

perior to the common "up country" rye; unacclimated, this last, after being grown here some three or four or more years, yields fair crops, and is sown to a small extent for bread; although otherwise valued by some, is inferior to the true Egyptian oats for winter pasture, or fodder. If desired, you can have a supply of these oats next summer for distribution from your office.

*Barley.*—Am not aware that it is grown. Have tried some half dozen sorts, as also *bear* or *big*, imported from Scotland, not worth the trouble and expense.

*Maize.*—Many varieties are grown, principally those known as flint and bastard flint. The gourd-seed varieties are very objectionable in this climate; principally on account of their softness rendering them unfit for bread, and open to the attacks of insects in the field and the crib. We require a grain white, hard, and rather flinty—white because of its great consumption in bread and hommony [i.e., hominy][11]; in the preparation of both of which our cooks greatly excel. When meal is ground for bread, the mill is set rather wide, that the flinty part of the grain may not be cut up too fine, this being sifted out for "small hommony"; the farinaceous part of the grain is left for bread. This hommony is a beautiful and delicious dish. On most plantations the negroes have it for supper, with molasses or butter-milk. A *hard flinty* grain is necessary to head the weevil, with which not only the cribs but the heads [ears] of corn in the field are infested. These are the *Calandra oryzæ,* the true rice weevil; distinguished from his European cousin by the two reddish spots on each wing-cover, and known among us as the "black weevil"; also a little brown insect, not a true weevil, but a *sylvanus,* as Dr. Harris writes me, to whom, through his invaluable work and private correspondence, I am indebted for much of that little I know of the insects injurious to agriculture. This sylvanus and another of the same genus, most probably the

being planted throughout that region. Affleck's assessment of this *Multicole,* likely based on his experience in the Lower South, does not match the enthusiasm that planters in the Upper South expressed for the crop.

11. Hominy is dried corn kernels treated with an alkali in order to kill the seed germ and preserve the kernel. The process, known as nixtamalization, was practiced by Native American residents of present-day Central America and the United States as early as 1500 B.C. and was adopted by early European settlers in the seventeenth century. Hominy, when ground fine, is masa, a staple of Central American cooking used as flour; when ground coarsely, it is either grits or corn meal, both staples of the cuisine of the American South. Then as now, crushed corn kernels are also used as livestock food.

*S. surinamensis,* affect the corn in the field before it becomes hard, causing serious damage—but nothing to equal that occasioned by the black weevil.

I know of no generally successful method of staying or even checking the injury caused by the insects; though much might be written in the way of suggestion.

Almost any variety of this grain planted even as late as the 10th of July, will ripen in our fine climate. There is thus no difficulty, where the land will bear it, of ripening two crops of corn on the same ground in one season.

As to the change of character and qualities from change of climate, I will speak anon.

I consider the shuck to be richer and stronger food than the blade,[12] which is but chaffy at best, when gathered from stalks which have matured grain, and is moreover the most costly article of fodder that is fed in any country. It costs some 8 to 10 per cent, of the grain, in weight and value, but being stripped before the grain is ripe. It costs no trifle from the injury the cotton crop sustains, from being deprived of a thorough working, at a stage when such a working is of great importance to that crop, by the necessity for fodder pulling at that very time. And the cost of pulling is great indeed, inasmuch as that a hand cannot pull, bundle and stack more than 3 to 400 lbs per day; during which his health suffers more than at any work; still, it is doubtful if the dependence upon blade fodder in the south be ever greatly lessened.

I suppose blade to be equal, pound for pound, to timothy hay as received here in bales, superior to crab-grass hay (*Digitaria sanguinalis*); all three inferior to sound shucks; and none of these at all to be compared with hay of Bermuda grass (*Cynodon dactylon*), the most productive and nutritious grass in hay or pasture of which I have any knowledge.

Of green corn, grown in drills[13] for stock of all kinds but hogs, and especially for cows and work oxen, it is difficult to estimate the value. I grow acres of it sowed in succession through the spring and summer, curing for winter fodder all that is not consumed when the ears begin to form. It seems wonderful how it can be dispensed with.

12. The "shuck" encases the cob; the "blade" is the leaf; vegetative matter such as the shuck and blade were used as animal food.

13. A "drill" is a shallow furrow in which seeds or bulbs are planted, usually created by dragging a hoe in a straight line, creating a shallow indentation in which seeds can be dropped, then covered. Planting by this method is more efficient than planting seeds individually.

When the work is done by plantation negroes, it is generally best to feed corn in the ear. Grinding and cooking are unquestionably economical practices; but difficult to be kept up with frequent changes of administration, under different overseers, where the planter is not always on the spot.

*Rice.*—I have grown, and some of my neighbors still grow, common upland rice with the most perfect success, gathering the heaviest crops known of any small grain; unless perhaps oats occasionally in Scotland. Grown in broad drills, say three or four feet apart, and tended with hoe and cultivator, requiring two good workings, nothing is wanting but some degree of encouragement, and the spreading of proper information, especially in the pine-woods regions, to make this a staple and a profitable crop. On the small pine-woods[14] farms in the interior of the State and along our sea coast [the Gulf Coast], upland rice is grown in considerable quantity and of fine quality; but for the want of mills or a market for the rough rice or *paddy* it is fed mainly to stock.[15] I forwarded a sample to Liverpool recently to learn its value as *paddy,* and have no doubt that it will bear shipping with profit. Such inquiries should be the duty of a general and state boards of agriculture. The tax upon the time and pocket of those individuals desirous of procuring and disseminating such information, and of introducing, testing and acclimating new trees and plants is too great, and should be borne by the general government.[16] The pine-woods farmers speak of their sandy lands, usually considered of little value, being capable of producing some three fair crops of corn, and four or five of rice, with an occasional crop of sweet potatoes, before they are utterly *worn out.* They also say, that if the straw be returned to the land it will produce many successive crops of rice. Manuring, except to a limited extent by cow penning, is never practiced.

14. Mississippi's "piney-woods" region, about a third of the state, is defined by an east-west line from Meridian to Vicksburg extending south to an east-west line slightly above the Gulf Coast. Gently rolling hills and sandy soils here would not support the kind of plantations found in the flat Delta region north of Vicksburg; instead, the piney-woods region was heavily forested with varieties of pine trees useful in the maritime, construction, and paper industries. Farms here were small, and often cattle were raised on land forested with pine trees.

15. Generally, "paddy" refers to the flooded fields in which rice is grown. Here the meaning refers to threshed but unmilled rice.

16. Here Affleck reiterates his belief that the national government should be asking agricultural questions and securing reliable answers, such as he reports here, in efforts to aid farmers and advance their economic interests.

In answer to your *note,* it would be the merest guess-work, equal to guessing at the growing cotton crop of 1st of September, to state the "cost of production," the "average per acre," or the actual aggregate product of our State; and therefore I will not attempt it.[17]

Of the "usual weight"—I had a good, sound common-sized flour barrel filled three times, settling the corn each time when filled, by shaking the barrel moderately, from the pile of corn (bastard flint, similar to the well-known Baden), as hauled and quite closely slip-shucked in gathering. When shucked and shelled, and the shelled grain poured into the barrel, it lacked four inches of being full, which I estimated at half a bushel.

The shelled corn weighed, net, 173 lbs. Add shattered and shelled off, where unsound at the points, 2–175 lbs., or 3 1/8 bushels of 56 lbs., our legal weight. The corn measure, in a sealed half bushel, 3 1/8 bushels. The corn was sound and good, with a moderate proportion of nubbins. It is to be inferred from this that a barrel of corn must be closely slip-shucked to average, to a certainty, a bushel of shelled corn; and our southern white corn, from which the blades have been stripped whilst yet green, will weigh, if sound, exactly the legal weight of 56 lbs. per bushel. All grains are bought and sold in the loosest possible manner, all through the South-west, unless in New Orleans.

The land being level, mowing machines might be used; more readily as the surface of a Bermuda meadow must be made very smooth before it can be cut to any advantage, even with the scythe. Hay, in New Orleans, is rarely so low as fifteen dollars, and is frequently up to $30 and even $40 per ton. Being on the river bank, the market could be watched and supplied when prices were highest, and there would be little or no expense of hauling. Land now rendered almost worthless by the bitter coco (cyperus)[18], may be applied to this purpose, as the Bermuda will overcome the coco, by top-dressing and mowing. I repeat (see Southern Agricultural Almanac of 1848, page 61) and can refer to numerous witnesses to prove, if needful, that we have measured the ground and weighed the well-cure hay, and this more than once, when *one cutting,* and that *the second one that season,* yielded over *five tons per acre.* After that, a very fair *third cut* was taken from the

17. As previously mentioned, these metrics were foreign concepts to most farmers, although Affleck's *Plantation Record and Account Book* would certainly have facilitated such analysis.

18. *Cyperus* is a large genus of sedges (grasses), many of which are invasive. Bitter coco (*Cyperus rotundus*) forms rhizomes that have a bitter taste and resemble nuts.

same ground. *Five tons per annum* is a moderate yield from a good, well-set Bermuda meadow, which is either top-dressed with sludge from an overflow or receives one of manure annually.

It affords equally valuable pasturage; but is a pest in the crop, only to be destroyed by a smothering crop of corn and pumpkins, clover or peas. By this means, I find no difficulty in *checking* and even eradicating it.

We have reports of hay made from *leersia orysoides*[19]—"Rice's cousin," as the negroes call it, and a valuable grass here, though pronounced by Dr. Darlington "worthless." From *Eleusine Indica,* or dogs-tail grass, "crow's foot," of this region, which on manured land grows with great vigor, though an annual grass. From Nimble Will (*Muhlenbergia diffusa*), which also, in some soils, and especially in woodland, originally of *Magnolia grandiflora* growth, but from which the magnolias have been cut, leaving only the deciduous trees, makes excellent pasture. And in wettish flat lands, from several varieties of *Panicum crus-galli,* which there grow vigorously, not infrequently mixed with *cyperus repens,* sweet coco, or nut grass. And some speak of hay from Guinea grass (*Tripsacum dactyloides*), which certainly grows vigorously, affording frequent cuttings, and objectionable only thus far, that in no condition or stage of growth can ever mules be induced to eat it freely; at least such is my experience.

I am not aware to what extent experiments have been tried with other grasses. I have imported from Europe seeds of over forty kinds, from Texas and the far West, over ten or a dozen, and have also tried any number of native (?) grasses with varied success, of which the relation might be of some interest; but will only remark here, that after careful and repeated trials, I have found no grass to compare, for hay or pasture, with the one commended above—Bermuda grass, the Doub or Dub, the sacred grass of the Hindoos. Of its value for summer grazing, I must state further that it far exceeds that of any other grass within my knowledge in abundant yield, in sweetness and in nutritive qualities. On the common around this village, there are cattle, horses, mules, sheep, goats, hogs and geese innumerable; all the year round, from the first evidence of renewed vegetation in the spring;

19. *Leersia orysoides* is rice cut-grass, a native wetland grass about two to four feet tall. Its seeds are attractive to waterfowl. Dr. William Darlington (1782–1863), born in Pennsylvania, served three terms in the U.S. House of Representatives. He established a natural history society in Pennsylvania in 1826 and published several works on botany and natural history and on the flora of Chester County, Pennsylvania.

and yet they are not all able to keep down this grass which covers the common; and during the summer, when it flourishes most, much of the stock is in fair order.

*Of Clover (see Plaster, &c.)*—It may be well to add that late in the fall, when the cotton is stripped of its foliage, the fields become green, where the soil is at all good, and various annual grasses and nutritive plants, which afford sweet pickings to stock, and especially sheep, all winter. There are the "winter-grass" of this region, the nearly universal *Poa annua*, here at times almost rank in its growth, reaching a height of from four to eight inches. Chick-weed (*stellaria media*), of which cows are very fond, as also sheep, covering the hill lands where rich with quite a heavy growth. *Phalaris Americana*, a beautiful southern grass depicted in Cellist's work. *Hordeum psuillum* of Nutt, a dwarf barley, or, as here called, "Texan Rye" forming sweet grazing before the blossom drops. *Alopicurus geniculatus*, floating fox-tail of the English, almost as valuable as the winter-grass. *Trichodium laxiflorum*, hair-grass, also springs up. There are nearly all annual *winter* and *early spring* grasses. In the fence corners may be found a good bite of Nimble Will, and on poor spots of habits like the Bermuda, has spread to a considerable extent over the open pastures. It is known by some as "Cuba-grass," and is a *paspalum* or *digitaria*, I know not which; the sheep find sweet picking from it. On the sea-coast, about Pass Christian and Pascagoula, I find a close good sod of another grass, of similar habit to the last named, of which I have not been able to determine the name.[20] It makes a very pretty pasture, and grows well even in a partial shade. Old pastures become infested with a coarse grass, growing in tufts, known as "Natchez grass," *agrostis Indica*, or black seed-grass. I think it of little value; in fact a filthy pest.

Such is an imperfect sketch of the grasses most common and useful in this portion of the south. It is a branch of botanical knowledge the most difficult to acquire, and assuredly sufficiently neglected. Would that the directors of the Smithsonian Institute might be induced to turn their attention to the subject, and give to the world a work upon the Graminæ [grasses] of this continent, native and introduced, worth of the subject.[21] If there is no hope

20. Pass Christian and Pascagoula are small communities on the Mississippi Gulf Coast.

21. The Smithsonian Institute was established in Washington, D.C., in 1846. In 1850 President Fillmore invited A. J. Downing to develop plans for its grounds and those of the adjacent U.S. Capitol. They were presented in 1851 but, upon Downing's subsequent death, were not installed.

of this, cannot your department take up the matter? There is no one topic of so much importance to the agricultural community. We have been again and again promised a work of the kind, but as yet nothing has appeared. Each and every grass should be depicted, and that in the very best style of the art.

I have said nothing of a grass frequently spoken of lately, "the Muskeete" or more properly "Mesquit" grass, and for the reasons that, though I have received, after much trouble and expense, various lots of "Muskeete grass-seed," comprising five distinct varieties, only one of them is of any value; and that I cannot name as yet, but will be glad to send dried specimens of this and all the other grasses to be found in this region, to two or more botanists, who can assure me that they have made this department their particular study, and who will aid me in identifying and describing them.

I am not aware that *irrigation* of meadows has been practiced.

*Peas.*—No varieties of the genus *Pisum* are grown, except in gardens.[22] But the pea, or more properly bean, known as the "cow" or "Carolina" pea,[23] is grown to a great extent, as food for man and beast, and for the improvement of the land. In all that has been written upon this very valuable plant, second only in value to maize in these Southern States, in no instance can I find any reference to its origin or botanic name. Having examined all the authorities within my reach, and caused many extensive libraries to be searched, and inquiries to be made in Europe, I have come to the conclusion that in any or all of its numberless varieties, it is hitherto undescribed, or described very imperfectly, and am therefore unable to answer the inquiry so often put, of "What is this cow-pea?" I am not competent to a botanic description; but will have pleasure in communicating the information acquired, and in forwarding seeds and dried specimens to botanists who are competent. It is evidently a *dolichos*[24]; but if described at all, is most probably classed as a *phaseolus.*[25] There are many *species* as well as *varieties*

22. *Pisum* is a genus of the family Fabaceae; three species are recognized, one of which is the pea.

23. Cow pea, black-eyed pea, or black-eyed bean are names for *Vigna unguiculata, V. sinensis,* and various subspecies. The plant, widely grown, is thought to have come to America from Africa.

24. A genus of flowering plants in the legume family, *Fabacae.*

25. This is the genus name for about fifty species of wild beans, cultivated in North America since pre-Columbian times. Some species originally in this genus have been transferred to *Vigna spp,* creating nomenclature confusion.

cultivated under the general name of *cow-pea;* ranging in size from that of a grain of wheat, to that of the smaller varieties of snapbeans. In color they vary still more; snow-white; white with black, red or yellow eyes; jet-black; purplish-red; yellow; speckled, like the early valentine bean, greenish-gray, like the gray field pea (*pisum*) of England, &c. Some grow very vigorously, covering the corn stalks, when planted among that crop, with a perfect load of vine and leaf; whilst others scarcely vine at all. The blossoms are of different colors and sizes in the several kinds; and the pods are some flat and some round; in some the pods stand out stiffly, in others hang loose; but all the kinds bear a strong family likeness.

The cow-pea is most commonly planted between the hills of corn, at the second hoeing. It does not vine much, nor bear pods until after the fodder is pulled; it then covers the stalk, ear and all, with a mass of foliage; affording, undoubtedly, a very large amount of food for stock, which are turned into the field after the corn is gathered, and vegetable matter to be returned to the soil. But the injury to the soundness and keeping quality of the corn, and the multiplication of weevil under the shuck from the shelter and moisture and soft condition of the grain—all of which weevil we carefully gather with the corn, and house with it in the crib—it is to be feared greatly counterbalance the advantages. Stock, too, are very frequently injured and in many cases killed by being turned into the cornfield to feed upon the peas. Many of the peas have sprouted or moulded, and not a few are in a state of partial decay before this can be done; hungry cattle and hogs are not very discriminating; hence the injury. Our southern agricultural papers contain many lengthy articles, pro and con. Although it is certain that great and sudden mortality has occurred among cattle and hogs, and occasionally even mules and horses, after having been some days in the pea-field, it is equally certain that a great majority of careful planters have been in the constant habit of consuming their peas in this way for many years without any such results.

It is, however, as a fodder crop and as an improver of the land that this plant is of the greatest value to the south. Land, when "turned out," that is, when so far exhausted by repeated croppings and ceaseless cultivation as to be no longer capable of yielding a remunerating crop—is generally so much worn out as to be unable to produce a crop of even weeds to afford protection from the sun. Even the cow-pea will not make a cover, unaided. Manuring, unless in some simple and easy way, will not soon be practiced, even to the extent to induce a growth of pea-vine. A cheap and easily applied manure for this purpose, I have found to exist in marl[26] and in plaster (sulphate of lime) and of which I shall speak under that head.

I can give but little idea of how many bushels of this pea is produced per acre; most probably, when grown amongst the corn, from ten to twenty. They have to be picked by hand, pod by pod; each pod contains from 18 to 22 peas. They sell at $1 per bushel.

*Root Crops.*—The Irish potato is grown to some extent, almost entirely for home consumption; unless near the rivers, where pretty large crops are occasionally grown for the New Orleans market. They produce well, and are large and mealy. Can give no particulars as to cost of production, &c.

The turnip is also grown in considerable quantity for plantation use; rarely for stock or for sale, though yielding large returns for the labor requisite.

Carrots, beets, mangold-wurtzel,[27] &c., only in gardens. The artichoke I have grown to some extent, but do not value it highly.

Skirving's improved Swedish turnip (Ruta-baga) I have found a very valuable root, productive and highly nutritious; and continue to grow them.

*Maranta arundinacea,*[28] which yields the arrowroot of commerce, I have tried so far as to prove that it may be made a profitable crop.

The "Pindar"[29] (ground-nut) is grown for market by the cultivator of sandy pine lands, and generally by the negroes; by some planters as food for their hogs, which are allowed to harvest them. I have found them an extremely exhausting crop for the land.

26. Marl is a loose soil (sand, silt, or clay) with large deposits of calcium carbonate.

27. Mangelwurzel or mangold wurzel is a cultivated root vegetable with large white, yellow, or orange-yellow swollen roots similar in appearance to a yam. It was developed in eighteenth-century Europe as a crop for livestock, although when young, both the root and its spinachlike leaves can be eaten by humans.

28. A perennial herb grown for the starch found in the rhizomes of its root system.

29. The "pindar" or peanut is a legume (not a nut) grown in the sandy soils of south Alabama and Georgia. As a legume, it fixes nitrogen in the soil, a beneficial soil-building characteristic that contradicts Affleck's assessment of the plant. *Pindar* and *goober* were words of African origin used for the plant by enslaved people. It was at the end of the nineteenth century that the research of scientist, botanist, and educator George Washington Carver (1864–1943) demonstrated the value of peanuts, soybeans, and sweet potatoes for southern farmers as alternative crops to cotton. Over his lifetime, Carver published forty-four agricultural bulletins for farmers, the most popular of which was *How to Grow the Peanut and 105 Ways of Preparing It for Human Consumption* (1916); by 1940 it had gone into seven editions.

The sweet potato is an important root; and they are grown in great quantities; though not to an extent commensurate with their value as an agreeable and nutritious article of food for man and beast. I have bestowed much attention on their cultivation; to the habit and growth, and to the comparative value and productiveness of the different varieties.

*Cotton.*—The questions under this head would require a lengthy treatise. You will, most probably, receive more than one essay in answer.

If you will send an artist this way, capable of making the necessary drawings and plans of gin-houses, presses, cotton-thrashers, &c. &c., I shall take pleasure in drawing up a lengthy article, which might, by such means, be rendered both interesting and instructive.

You will perceive that in the *Plantation Record and Account Books* sent herewith, I have provided for much of the information sought for; average yield per acre and per hand; cost per pound or per bale, or production freight charges, commission, &c., paid by planter, and more of a similar character and of like importance.

*Sugar* must be left to those having a better knowledge of the subject. It has within the last three years been grown successfully and profitably in the hills thus far north, and upon lands which no longer produced remunerative crops of cotton. There is no reason why it should not displace cotton to a great extent. P. M. Lapice, an extensive and very enterprising public-spirited planter, in the Parish of St. James, has demonstrated for some fifteen years or more, that sugar-cane thrives as well, and ripens as many joints, at his cotton plantation opposite Natchez as in St. James.

*Hemp* has been grown in the State on the banks of the Mississippi, and that successfully, but I have no knowledge of its cultivation as a crop.

*Butter and Cheese.*—The former is made in as great perfection as in any part of the world. Every planter's wife makes an ample supply for her table, and occasionally enough for the inhabitants of the towns in the vicinity. But, as a business, I am not aware that it is carried on within the limits of our State; although there is no part of the Union where it could be made so profitable. Good butter averages the year round of 25 cents per pound, often commanding 40 cents, and never under 25 cents.

*Land* is cheap. Good cows can be had at moderate prices from $15 to $40, yielding from two to twenty quarts per day, according to the selection of animals, and the manner they are fed. With industry and judicious management, abundant pasturage and an ample supply of green food can be had all the year round. During continued wet weather in winter and early spring, when it would not be advisable to allow the cattle to puddle the land, fodder from the cow-pea, cured vines, peas, and all, as is commonly practiced, with Swedish turnips, beets, carrots, sweet potatoes, cabbages &c., may be used. A farm should be properly arranged for the business, subdivided, that separate pasture lots of Bermuda grass might be grazed alternately from the first or middle of May until December; buildings erected, or a gin-house altered to serve the cows; cisterns for water, and tanks for liquid manure, to be applied to the grass land kept for the scythe. The subdivisions may be effectively and cheaply made by means of the Cherokee rose.[30]

For the planting and cultivation of this plant see Nos. 1 and 2 of vol. 5 of De Bow's Review.[31] When a good pasture of Bermuda grass is kept ungrazed, in the fall, so that the grass grows to a height of six or eight inches, the early frosts do not injure it so far as to prevent cattle from getting a good bit until mid-winter. Clover, or Egyptian oats, or rye, sowed in September, may be grazed after Christmas; at intervals, the oats until the 1st of April; the clover until June. If soiling were practiced, all the liquid manure saved in tanks, diluted and applied to Bermuda meadows, clover, peas, drilled corn, &c., from watering carts, the improvement of the land would be rapid and the yield of fodder immense. It is almost impossible for the cotton planter to carry on anything of this kind. It could only be done to advantage on a regular dairy farm.

*Horses and Mules* are now bred in considerable numbers in some part of the country, and many of them splendid animals. The business is found to

30. "Liquid manure" was a concoction (often called "tea") made by combining manure from horses or cattle with water and letting it steep for a period of time, then applied directly to the roots of plants as an effective means of fertilizing. Affleck often wrote about using the native Cherokee rose as a means of marking property or enclosing fields. See his discussion in the section on roses that follows.

31. *De Bow's Review* was a widely circulated journal of "agricultural, commercial, and industrial progress and resource" published in the South from 1846 until 1884. Its founding editor, James Dunwoody Brownson De Bow (1820–67), wrote many articles in the early issues and corresponded with Affleck. For most of its run, the paper was printed in New Orleans, but its sphere of influence covered the entire South, if not also the nation. In its early years, the paper was designed for a broad audience and it printed articles on agriculture, economics, literature, political opinion, domestic matters, and editorials. Unlike contemporary agricultural papers, *De Bow's Review* took an increasingly pro-secessionist position, advocating southern nationalism as the Civil War approached. After the Civil War and following De Bow's death in 1867, under different leadership the journal took a different position, urging the acceptance of Reconstruction.

be profitable, and does not in the least interfere with the cultivation of either sugar or cotton. Animals bred here are much more hardy and durable than those brought down the river.

It seems unaccountable that planters in Mississippi do not set determinately to work to render themselves independent of their neighbors for the supply of an item so costly. If they were even to purchase yearling mules from the breeders in Tennessee, Kentucky, Illinois, Indiana, Missouri, and bring them south at that age, they would find the business still more profitable than it has long proved to be to the graziers of those States, who regularly buy up the young stock from the breeders. They are bought when weaned at from $20 to $30; the cost of transportation by steam would be less at that age than when grown, not exceeding $5 per head, including feed and insurance when a number are shipped at once; the cost of keeping on a plantation for two years would scarcely be felt; whilst the mules would be worth in reality, one half more than if bought south at three years old. Some planters breed the largest sized "cane-tackeys," as they are called, or native ponies of Spanish origin, to well-bred but small stout horses; thus producing a stock of tough serviceable animals, almost as durable as mules.

Mules vary in price from $70 to $125. Horses from $50 to $300. Cane-tackeys, good stout ponies of 12 to 14 hands, from $20 to $50.

*Horned Cattle.*—I have no means of answering your inquiry as to the number in our State. It is immense, however, and especially in the interior. The principal markets are New Orleans and Mobile, with the smaller towns in the state. A large number are annually consumed in the teams from the state of the roads during the hauling season, the carelessness of planters and overseers, and the cruelty and rascality of negroes—who often drive their beasts for days with scarcely any feed, reserving what was given them for use on the road to sell for their own benefit.

The native stock of the country, the large brick-colored or brown oxen, with their singularly twisted ram-like horns, are an excellent breed, making noble teams and fattening readily. Some of the cows, too, are fair milkers. Herefords, Durhams and Ayrshires, have been introduced in considerable numbers and at great expense; but from various causes, have done but little good; the Durhams least of all. Some of their crosses on good native stock are fine animals.

The average price received by the breeder for three-year old unbroken steers, sold to the butcher or drover, is from $6 to $15, according to condi-

tion and locality, and at these prices they pay pretty well. Good well-broke teams are worth per yoke from $35 to $60. Cost of keep, I have no means of estimating, nor is it ever taken into consideration.

*Sheep Husbandry* is a subject of decided interest to this State. It would, however, require a volume to answer the questions propounded as they should be answered.

Mr. Randall, in his letters to Mr. Allston, has accumulated a mass of information greatly encouraging to those who have the desire to engage in this business in the South, lacking, however, much that experience in a Southern clime alone can give. I will only remark that the short and fine-wooled families of sheep do well, whilst the long-wooled do not. I am not aware of a single instance in the South-west in which Cotswolds or Bakewells have done more than to exist for a year or two. Some of their crosses upon the shorter-wooled kinds have done better.

The Southdowns succeed admirably, and have greatly improved the mutton of the country, so far as *fatness* is concerned; as to fineness and flavor, doubtful. The Merino, Saxony, and Saxony-Merino thrive well, and in my opinion improve in quantity and quality of wool.

The improvement effected by a first cross upon our native ewes is great and uniform, both in wool and mutton. I have a small flock of very superior animals, Saxony and Merino, brought South four years ago. During the first two years they throve very badly, and increased slowly. Now, however, they do well and breed freely. I have compared the wool carefully, each clip, and think I see a marked improvement in fineness and softness. As yet I am not prepared to say more. Few planters keep more sheep than enough to supply their own tables with that most excellent dish, a saddle of Mississippi mutton, which compares favorably with the mountain mutton of Scotland and Wales. They suffer at times severely from dogs.

*Hogs.*—The queries under this head must also be passed over, for reasons similar to those given above. Many planters raise an ample supply of hogs for their families, black and white. Many more find it a thing impossible, from the destruction of their young stock by the negroes, who have all a particular penchant for roast pig, and especially when stolen; and many never make an attempt to raise pork.

*Rain.*—Herewith you will receive an almanac, in which is a table of the temperature, quantity of rain, &c., which I condensed from the register of the late Dr. Tooley, of Natchez, published in *Silliman's Journal.*[32]

*Labor, its Cost, &c.*—Negroes hire out readily at $15 per month for common out-door labor; the owner clothing them, paying physicians' bills, if any, taxes, &c.; the employer boarding them. When hired by the year on plantations, which is rarely done, the employer pays about $70 to $75 for a full hand, paying all expenses, in sickness and in health, unless perhaps taxes, and supporting the children if any.

White laborers, when making levees, canals, ditches, &c. receive $1 per day and board with *quan. suff.* [quantity sufficient] of whisky. Few owners will put their negroes at such work in the swamps, mainly on account of its unhealthiness.[33] At work in the mills they have from $10 to $15 per month and board, Carpenters $30 to $50. Gardeners from $20 to $50. Overseers $250 to $800, according to number of hands on the place, and the experience and competency of the overseer.

---

32. The *American Journal of Science and Art* began in 1818, and its editor was Professor Benjamin Silliman. Still published, it is America's longest-running scientific journal. Its current focus is on the natural sciences.

33. The practice of using white labor—mainly immigrants—for hazardous urban infrastructure tasks in the American South (and elsewhere) was not uncommon. In the 1840s, streets and drainage systems in New Orleans were built by newly arrived Irish and German immigrants. Many died from malaria, yellow fever, snakes, or alligators while constructing the New Basin Canal in the swamps just west of the city.

The number who have gone, or are going to California, has somewhat raised the wages of overseers. Intelligent young men from the North and West, who are pretty good farmers, would find employment in this capacity; being content with moderate wages for a couple of years, under the eye of experience planters on their home places.

*Tar and Turpentine.*—In the almanac already referred to you, you will find all the information I possess on this subject. A number are engaged in the business and find it very profitable. There is great natural wealth locked up in Mississippi, from the want of a complete survey and report, geological, agricultural, and economical.[34]

*Plaster*[35] *and other Fertilizers.* During 1842, '43, and '44, I tried repeated experiments with red clover, sowing at various times, but mostly in the winter and spring. The result was invariably the same; so soon as warm, dry weather set in, the plants, though previously making a fine growth *on a good soil,* began to wilt as the day advanced, and by evening were entirely wilted down. By the first of August, scarcely a plant was to be seen, except in the fence corners and around stumps. In September, 1845, I broke up about an acre, consisting of three or four sharp and steep points of *very poor land,* with hollows between, of *good rich soil.* The hollows had been thickets of brier; upon the points scarcely even a stalk of broom-sedge or of poverty-grass could exist. I sowed the whole immediately with red clover, giving the poor portions a powdering of *plaster,* so soon as the seed had sprouted, at the rate of 1½ bushels per acre. During the fall and winter, the clover grew vigorously, showing little difference between the points and the hollows.

As the spring advanced, the plastered portions assumed a deeper green, never flagging a leaf during a very dry time. This induced me to have a bushel of plaster cast over the whole, the poor land thus receiving a more than double portion. After the very first shower, the effect was manifest and great. The clover on the whole of the lot grew vigorously through the summer, *the poor land keeping the lead.* I cut the whole over the soiling and for hay, when in full bloom about the first of June, I think, getting heavy cut.

34. Turpentine comes from the sap of pine trees, notably longleaf (*P. palustris*) and loblolly (*P. taeda*), which are found abundantly in the "piney-woods" region of Mississippi. Turpentine and other pine-related products became an important source of economic activity in south Mississippi in the late nineteenth century and into the twentieth century.

35. The reference here is to gypsum, a common sulphate mineral often known as plaster of Paris, which can be used as a fertilizer.

It grew out again vigorously, bloomed and ripened a fall crop of seed, and, with the exception of a few straggling plants, died down, roots and all. About the middle of September, the seeds sprouted, and by the middle of October, the lot was as green as ever. It now got another bushel and a half of plaster over the whole. The weeds and briers had been kept down. By the middle of May it was a rich sight, the clover standing fully as high as the knee of a tall man, and covering the ground thickly and evenly, but most so *on the richer land.* It was again cut, the clover being removed only from the richer spots. On the rest, it was evenly spread and left as a top-dressing. The second crop was much better than the first upon the ridges, showing distinctly the effects of the top-dressing. The whole was cut *when in blossom,* and made into hay, *which salivated* [i.e., gave black patch disease to] *everything that ate it,* horses, mules and cows, and was ultimately used for bedding. The first crop did not produce this effect. The second crop salivated even hogs, turned upon it to graze. The third year the clover was not so good, still yielding fair crops, however, of which the first was made into hay, and the second ploughed in when at its best. Last year it was in corn, without manure of any kind, and was 50 per cent. [This is] better than on similar land, differently treated. Even the poor ridges have a full crop of large, well-filled ears. This year I am appropriating it to a permanent layering-ground for evergreens, being within the limits of my nursery. It faces the north.

The results of other experiments, upon land of different degrees of richness, and with every exposure, have been the same; proving distinctly that red clover can be grown as successfully here as in New York or Maryland, when *manured with plaster;* that the plant becomes here almost an annual, but few continuing to live after going to seed in the fall, and these being weakly when compared with seedlings. Of many neighbors who saw the results of these experiments, and had the matter explained to them, but one has followed them up, and *that a lady!*

With the cow-pea in all of its varieties, with vetches, tares, lentils, the different garden beans, young locust, acacias, &c., white clover, the results are equally marked. By the application of from 1½ to 3 bushels of ground plaster to the acre, a heavy cover of peas can be produced upon the poorest lands of this region. The peas ploughed in and followed by clover, or Egyptian oats and clover, and then fed off in the spring to be followed by peas, also to be eaten down by hogs and sheep, sowed again with clover in the fall, to be carefully turned under in the spring, will renew any land.

Wherever plaster is applied to land not utterly worn out, a thick cover is produced of white clover, or a rank-growing species of *medicago* or snail-clover with a small yellow blossom, relished only by cows and some mules.[36] The white clover salivates every kind of stock so dreadfully that I look on it as a pest.

I have also applied plaster to Bermuda-grass and to corn, but not with the same effect as that produced on clover, &c., still decidedly beneficial.

I have procured the plaster used partly in New Orleans, at from $1.25 to $1.75 per cask of about 6 bushels, the freight and drayage being another dollar; and partly from the makers of soda water.[37]

Eight years ago I picked up some crystals of pure sulphate of lime, in the form known to geologists as *selenite* [a variety of gypsum], near Clinton, in that State; they were exposed in excavating for the Vicksburg and Jackson railroad. On farther investigation, I found it in considerable abundance, and have no doubt that an ample supply could be there obtained. I enclose you a specimen. About the same time, I pointed out extensive and inexhaustible beds of marl, in many parts of this district, the existence of which was hitherto denied here. I have used it to some extent and with marked advantage. Herewith you have a specimen, which please hand to my friend Dr. Gale for analysis, the result of which I should be glad you would add in a note. From a partial analysis made, it is rich in carbonate, phosphate and sulphate of lime; its effect upon the land is decidedly mechanical and chemical; making it friable and easy to work, and retentive of moisture; whilst it furnishes much that our soil and subsoil require of inorganic matter. It would occupy too much time and space to specify the results of different experiments. Suffice it to say, that in her vast beds of rich marl, Mississippi possesses a means of improving all of her worn-out lands that are not already too much gullied over to be reclaimed.

*Guano*,[38] upon many plants; the result upon a part of the sweet potato crop this season was most marked. The soil thin, worn, yellow clay; the po-

36. *Medicago* is a genus of low, flowering legumes resembling clover, commonly known as medick or burclover. The genus includes eighty-three species, the most common of which is alfalfa (*M. sativa*).

37. Gypsum is a by-product of making soda water.

38. Guano is the excrement of seabirds, bats, and seals. Found in great quantities on islands off the coast of Peru, it is an effective fertilizer due to its high concentration of phosphorus and nitrogen, elements which substantially increase the productivity of soils deficient in organic

tato variety *yam;* the guano at the rate of about a bushel per acre mixed with an equal measure of plaster sown along the ridges, the plough immediately following; the result, *a less growth of vine,* with more than double the quantity of potatoes, and these are large and fine.

*Lime,* only used as first related in the shape of marl, or of sulphate of lime and on fruit trees, as will presently be stated.

*Orchards.*—No portion of the Union is blessed with a soil or climate more favorable to the production of fine fruit than this and most other parts of these Southern States. An opposite opinion has unquestionably, but most erroneously, been entertained. There have been many failures certainly, but from very obvious causes; whilst many, the writer being of the number, have succeeded in ultimately overcoming the difficulties to be contended with, and in producing fair crops of fine fruit, apples and pears included.

The greatest impediment to success has been *the want of acclimated or naturalized trees* to begin with. Of peaches, such were to be had, native seedlings many of them decidedly fine. But for apples and pears, the sole resources were the nurseries in the Northern States and Europe.

The subject of the acclimation of plants has been one of long and earnest dispute among the learned. As to their theories, they are of no moment; the facts are these:—

Every planter knows that if the corn in the crib be likely to prove short he may secure a supply, at least six weeks before his main crop will ripen, by planting a few acres of *boat corn,* that is, corn the production of a more northern climate, most commonly of Ohio or Kentucky.

That the yield, though earlier ripe, will be lighter than if he had planted seed grown here, that if the second year he plants of the produce of his *boat corn,* the plant will be *later, stouter,* and *more productive,* though not so early

---

material. During the 1840s and 1850s, agricultural papers published many articles about the efficacy and application of guano, and speculators rushed to bring large quantities back from Peru for distribution in the United States. With its high concentration of nitrates, guano also became a key ingredient in gunpowder, and this was another reason for its importance in the mid-nineteenth century. Because of its widespread use, the U.S. Congress in 1856 passed the Guano Island Act that enabled U.S. citizens to take possession of any island, anywhere, that contained guano deposits, so long as it was not occupied or under the jurisdiction of any other government. Guano remained agriculturally important throughout the nineteenth century, but its importance diminished significantly after 1909, when an industrial method of fixing nitrogen enabled an efficient way to produce fertilizer chemically.

by some weeks as the year before, and that a third or a fourth year identifies it with southern corn. We may add the singular fact, that our native or southern corn will stand uninjured by a late spring frost, which shall cut to the ground and utterly destroy the plant from northern seed, growing in alternate rows!

The fact is notorious, that though garden seeds of northern growth will generally give us earlier vegetables, snap beans become stringy, squashes hard, cucumbers ripen, lettuce and cabbage go to seed, &c., much sooner than when from southern seed.

With roses and other shrubs, it is necessary to get a new growth *from the root,* entirely new wood, before they will flourish or thrive well.

*And so it is with fruit trees.*—The wood grown in a cold climate is adapted to that climate—to a cold long winter, and a short summer. When such trees are brought here, even if received in good order, which is a rare circumstance, they may grow off with some appearance of vigor for a time; but when warm weather has set fairly in, they begin to suffer; the leaves look dry and shriveled up; the tree is with difficulty kept alive by wrapping the stem and branches with moss, by mulching and watering, but soon, most commonly by the middle of July, every leaf has dropped and the tree either dies outright, or lives to make a faint attempt at a second year's growth. Occasionally, when headed well back, that is, the branches shortened, or the stem cut down to within a short distance of the bud or graft, and treated as just mentioned, they may live and grow, and ultimately bear fruit, but sparingly.

In the meantime, by budding upon vigorous native stocks, one step towards acclimation is made. In no instance have I found an old stem, wood or even one year, overcome the effects of this change of climate, and make a thrifty, vigorous fruitful tree.

Of three hundred varieties of apples, two hundred of pears, thirty of cherries, forty of plums and now in cultivation here at Ingleside,[39] fully one half will ultimately overcome the effects of change of climate and become naturalized. Every means is being employed to bring about so desirable an end.

Pears, which seem to adapt themselves of all fruits the most readily to our climate, are being grown upon several varieties of the quince, some

39. Recall that Ingleside was Anna Dunbar Smith's plantation upon the death of her first husband and prior to her marriage to Affleck. It became the site of Affleck's Southern Nurseries, and it was here that Affleck held plants to acclimate them to local conditions. This practice, followed by few other suppliers, made horticultural sense but was not financially profitable.

upon their roots or upon seedling pears; and others, such as the sekel, on apple stocks; or the St. Michael, on the French Doucin apple.

Of apples, some are worked in the usual way on seedlings; others on their own roots, and on the Paradise or Doucin apple.[40] The cherry thus far grows vigorously, the leaves persisting until frost on the mahaleb or perfumed cherry. The *monstreuse de Negel* and several others promise fruit next season, and so of other fruits. The quince as a stock for the pear, the Paradise and the Doucin for the apple, and the *cerasus mahaleb*[41] for the cherry, all have the same effect; that of dwarfing, and causing early maturity and fruitfulness.

My experience leads me to state emphatically, and that of hundreds of others will bear me out, that success in fruit-growing, thus far south, need not be expected where reliance is placed on individual trees of northern growth, though, by perseverance and some degree of skill, most of the finest northern and European *fruits* may be successfully naturalized.

Another important item to be considered, in growing fruit this far south, is the protection of the stem and main branches of the tree, and the shading of the soil around the roots from the powerful rays of the sun; to be properly effected by training the trees with a low head, and, at the same time, encouraging a thrifty growth; thus insuring an ample foliage. All trees seek to protect themselves in this way; and especially those with a smooth, glossy bark, which is so well calculated to absorb the heat of the sun's rays.

The bark of an apple, pear, or peach-tree—upon which the bright sun of this climate has been pouring his rays, through a long summer day, is *hot* to the touch, and even the sap will be found, on applying the tongue, to be of an equally high temperature. How can healthy, unblemished fruit be expected under such circumstances?

It is of more importance here than in a cooler climate, that the point of junction between the stock and scion or bud be *at* or near the ground. The causes need not be stated—to every nurseryman they will be obvious.

Overgrown, forced trees, produced by very rich soil or heavy manuring, do not suit a southern climate, nor, in fact, are they anywhere equal to those of a moderate growth.

Especially are they objectionable for planting out on the poorest hill lands of the south. In such a location, trees which have been grown on rich

40. Both of these varieties are ancient, brought to America from Europe.
41. *Cerasus mahaleb* is a cherry native to Central Europe and Asia.

land *starve* and burn out directly, even if well manured, whilst thrifty trees of moderate growth, whose shoots are short-jointed and well ripened, scarce receive a check.

The difficulties referred to in procuring acclimated trees of pears and apples, have led to the almost exclusive cultivation of the peach, which though occasionally produced of the highest excellence and in great abundance, is an uncertain crop; not only so, but as a fruit for market, they are altogether inferior to the apple and especially the pear.

Great care is requisite to carry them in safety to New Orleans; the more, as freestone peaches alone are saleable in that market, whilst an accidental delay of a day or two entails an entire loss. The pear and the apple, on the other hand, very rarely fail of a good crop; may not only be carried to New Orleans in perfect safety, and in good order, but will suffer no injury from a detention of several days; and their season extends through some months earlier and later than the peach.

As it is my intention to give to the world, in a few months, a familiar treatise embracing the *entire gardens and orchards in the South,* I shall not extend these remarks much farther.[42]

The low lands of the Mississippi, where dry or properly drained, are admirably adapted to the growth of the pear; and more especially when worked upon those free-growing varieties of the quince used for this purpose.[43]

It has been stated that the fall and winter varieties of the apple, those that are such in a northern climate, are worthless here. This is altogether a mistake, and has arisen from a misunderstanding of the matter. Although the summer and early fall sorts are most profitable and uniformly productive here, the latter kinds are almost equally so, but ripen at too early a period to keep well, or in fact to keep for any great length of time after they ripen, say more than a month or six weeks; and, indeed, I have found that the later the period of ripening the more difficulty in acclimation. Still many late fall and winter fruits, the Newtown pippins, for instance, produce and thrive well, and by perseverance I hope in a few years to succeed equally well with

42. While such a treatise did not appear, Affleck's annual *Almanacs* from the 1850s contain much information on these topics.

43. This pear is likely the sand pear, *Pyrus pyrifolia.* It produces round to oblong fruit that remains crisp with a sandy texture and mild flavor, not unlike a quince. More a cooking than a table pear, it can be stored without spoiling. It is commonly found today around abandoned rural homesteads.

a great proportion of the finest kinds. Ripening, as all the sorts do, long before they can be brought down the river, they command high prices.

*Lime is an absolutely indispensable ingredient* in the soil in which fruit trees of any kind are grown, and especially the apple and pear.

Until I was convinced of this fact, I found great difficulty in producing a healthy and vigorous growth upon many varieties of the apple. By marling, I removed the difficulty; the wood became short-jointed and healthy, the foliage abundant and persisting until frost, and the fruit large, sound, and free from specks or blemishes, such as before disfigured some kinds.

On grapes and the making of wine, it is unnecessary to say more than reiterate the statements of Dr. Weller, of North Carolina, relative to the incalculable value of the *white Scuppernong* to these Southern States for this purpose, and for the table. It succeeds fully as well here as in North Carolina, whilst the fruit is decidedly larger and the juice richer. It is a native grape, bearing the same relation to the muscadine of the woods that the Newtown pippin does to a crab-apple.

I have thus endeavored to answer your several queries as fully as possible, at the same time condensing as much as practicable.

I am, dear sir,

Yours with respect and esteem,

THOMAS AFFLECK.

Hon. Thos. Ewbank
*Commissioner of Patents.*

# 6

## Affleck in New Orleans

### Daily Picayune *Articles*

Upon moving to the Natchez area, Affleck must have soon realized that New Orleans was the commercial center of the region, and that success in his new nursery enterprise and continued visibility as a journalist depended on the connections he could make there with clients and leaders in businesses related to plants and publications. His surviving commercial records document lengthy correspondence and business relationships in New Orleans.[1] From the 1840s until Affleck moved to Texas in the mid-1850s, much of his writing was dispatched from Washington but appeared in New Orleans publications and was intended for audiences in both locations. For instance, Affleck's *Almanac* had similar instructions for Natchez and New Orleans, adjusted for differences in growing conditions.

Among Affleck's many contacts in New Orleans were "seedsman and florist" F. D. Gay[2]; J. D. B. De Bow, publisher of the influential *De Bow's Review*; and Benjamin Moore Norman, who published an early guide to New Orleans (1845), the iconic 1858 "Norman's Chart of the Lower Mississippi by A. Persac" map, and the *Norman's Southern Agricultural Almanac*. Affleck edited the 1847 edition, and this publication evolved into *Affleck's Southern Rural Almanac*, which appeared through the 1850s and into the early 1860s. Norman also published Affleck's *Plantation Record and Account Book* for sugar and cotton plantations in the 1850s, which were distributed nationwide.

In addition to turning out the almanacs and plantation account books, Affleck also wrote for at least three New Orleans newspapers: the *Daily Picayune*, for which he was agricultural editor in 1851; the *New Orleans Commercial Times* (1846–52); and the *New Orleans Price-Current and Commercial Intelligencer and Merchants' Transcript* (1851). Locating copies of these

1. For more information on Affleck's business connections in New Orleans, see Douglas, *Public Spaces, Private Gardens,* 191–95.

2. As described in his newspaper advertisements.

newspapers has proved difficult, but three articles from the *Daily Picayune* follow, covering a range of topics: brief mention of crops in the Natchez area, together with the suggestion of the community's attractiveness; a well-reasoned discussion of improving urban drainage in New Orleans; and an article on the fuchsia, "that most beautiful parlor flower, most admired by the ladies of all Flora's gems."

Particularly noteworthy among these articles is Affleck's assessment of drainage problems in New Orleans, including soil subsidence, saturated soil conditions and their impact on gardens, the lack of shade trees to deflect the heat, and the general lack of effective drainage. Affleck states that his views are those of a "farmer, not a professional engineer" and, when applied to a large scale, "may be open to some objections." Nevertheless, his proposed solutions—including underground drainage pipes and the use of run-off to irrigate fields for cattle—indicate the facility with which he could apply his experience on the farm to urban conditions and echo water management ideas discussed today. All three topics have interest for today's students of the designed environment, and they show that his range as a journalist matched his interests in agricultural matters and talents for creative solutions.

—Ed.

## The Crops in Mississippi

New Orleans *Daily Picayune*
The Crops in Mississippi, &c.
Washington, Miss., July 14, 1849

*Editors, Picayune*—As this "state of the crops" forms an interesting item of news to a great majority of your readers, it is well to keep you advised on the subject.

It is now four weeks to-day since these rains set in, and during that time there have been but three days during which it did not rain heavily; and for six weeks past the weather has been dark and showery. Grassy crops, in the extreme, and long-jointed, weeded cotton are the results. The rains, until from two to three weeks ago, were partial; so that crops suffered in some neighborhoods, and especially the early plant crop. As a general thing, however, cold will be abundant. It is time that we were pulling fodder; but in the present state of the weather—for there is not a symptom of its clearing up—that is out of the question.

In this district the prospects for an average crop of cotton are by no means good; and should the cottonworm make its appearance this season—and it is a most favorable one for its increase, the plant being rank enough to *produce it* or anything else but cotton!—at the usual time, the result must be disastrous in the extreme.

It is "passing strange" that a great number of planters in these healthy hill districts do not lend their attention to growing fruit—pears, apples and peaches—for your market. All three thrive well, but especially the pear and apple. The idea has prevailed that neither of these fruits can be grown this far south, which is altogether a mistake. As fine crops of both—limited crops, of course, as yet, the trees being almost all young—are grown every year, in this vicinity, as anywhere in the world. Acclimated trees, and these supplied with lime or marl in the soil, and the trees reasonably well tended and trained low for the protection of their stems from the sun, are the only requisites. There are several nurseries now amongst us, where acclimated trees can be had; and marl is abundant through all this region.

The health of Natchez and the region around is excellent. Our little village of Washington is a favored spot in this respect. This fact, together with the beauty of the location, the excellence of our schools and of the society in and around the village, is drawing numbers here for summer residences.

The Elizabeth (female) Academy, where so great a majority of the mothers of Adams and the adjoining counties have been educated, continues to sustain its long established position and reputation. It has now been for many years in the hands of Mr. and Mrs. Forde, and their accomplished daughters, giving entire satisfaction to parents. When young girls are sent from under the mother's eye for their education it is of the very first importance that she in whose hands they are placed should have proved herself, by the results in the training of her own daughters, *a mother in deed and in truth.*

"Old Jefferson" is to be reopened in Monday next, as a high school by Professor Haderman, well known as a teacher of the right kind for our Southern boys—firm but conciliatory in his discipline, making himself the friend and confident of his pupils; an admirable modern linguist and mathematician himself, whilst he has engaged able assistance in the other departments.

It seems to me that for those of your citizens who have families to educate, "our village" would be a much more pleasant place to spend the summer in than any of the watering-places across the lake.[3] The expense would be much less—the location healthier and more retired—no risk of being caught amidst such crowds by yellow jack or the cholera—fruit abundant, socially good, schools unsurpassed, &c. &c.

Yours, truly and respectfully,

THOMAS AFFLECK.

———◆———

## Drainage of the City

New Orleans *Daily Picayune*
April 26, 1850

Drainage of the City.
Washington, Miss., April 19, 1850.

Having been favored with a copy of the annual Report of the Board of Health of New Orleans for 1849, with the "compliments of the author," I would beg a small space in your columns to tender my thanks to Dr. Bar-

---

3. The "watering-places across the lake" are the small communities (Covington, Abita Springs, Mandeville) across Lake Pontchartrain from New Orleans that were popular summer retreats from the heat and diseases (yellow fever and cholera) of urban New Orleans. Access to these communities was by boats that regularly sailed from the lakefront across Lake Pontchartrain.

ton, and to offer a few hurried remarks anent some of the views advanced by him.

This report is an extremely interesting one, and conveys a knowledge of many important facts, with suggestions equally important.

The leading causes for the insalubrity of the city—the extent of which insalubrity, by the way, has been greatly exaggerated abroad—are justly stated to be great elevation of temperature[,] (want of free) ventilation, undue moisture, and filth.

This undue moisture, the absolute saturation of the ground, is literally at the foundation of the evil. And having made this the subject of frequent conversation with yourselves, your worthy Recorder and other men of intelligence in your city, with suggestions for remedying the evil, I desire to repeat them here. They are the view of a farmer, not of a professional engineer; and as applied to operations on so large a scale, may be open to some objection, though the principle be correct.

1st. Then, as to the necessity for and the advantages of thorough drainage, and especially in your city—

As is shown in the report before us, the health of the city is very injuriously affected by the saturated condition of the soil. The value of property is lessened and the expense of building greatly increased. Witness the settling of most of the large buildings notwithstanding the expansive foundations laid down; the injury to goods, of almost every kind, from the damp; the condition of our streets, [in] spite of the labor and outlay bestowed on them, even in such minor matters as the laying of a ground floor, of a gas or water pipe, the expense is greatly enhanced. Trees will not grow, gardens scarcely deserve the name, although extraordinary exertions are everywhere made, ending in disappointment and loss. Dr. Barton remarks that much of the insalubrity of the city arises from the elevation of temperature, and suggests shade trees and verandahs as remedies. The first is decidedly the best, but without thorough drainage to the depth of at least four feet, altogether unattainable unless by planting those few trees which grow in wet places. I have observed, in the back parts of the city, on the banks of the canal and of certain large ditches, [illegible] vigorous looking trees.[4] Trees, too, in addition to the shade they afford, are important purifiers of the air.

4. Here Affleck is referring to an area above the Vieux Carré known as "backatown." Located nearby was the Carondelet Canal, which connected Bayou St. John with the rear of the Vieux Carré. It was dug in the late 1700s and filled in the 1930s.

Again, the ground is not only saturated with water, but with many gases offensive and injurious to health, from which, if thoroughly drained, it would be gradually but certainly relieved.

When conversing on this subject, I am almost always met with the objections that there is no outlet for the water, nor fall sufficient for drainage, and that it would be impossible to lay drains, from the quantity of water which flows into a drain of a single foot or two in depth.

Dr. Barton states the "difference of level between Levee street and the (late) swamp beyond Broad street is about eleven feet, the basin and canals of the draining company are seven feet lower, making eighteen feet."[5] The consequence is, that underground drainage could be made from about Levee street, letting in the Mississippi river at ten to eleven feet below high water mark, by drains constantly open, which would produce a current to Roman street, the distance of about a mile, more than twice as rapid as the Mississippi at high water, and from Roman street to the basin of the draining company, more than three times as rapid, and of course three times as strong; a force amply sufficient to keep itself perfectly clear, and remove all filth and offal of the city.

The difficulty as to the superabundant water is met by beginning to cut the drains, of course, at their outlet.

The main sewers in the city of Glasgow, Scotland, where stone is abundant and cheap, and cheaply laid, have all been recently, I understand, made anew of cast iron, as the cheapest and most effective lasting material. Being cast in such a way as that the different sections fit into each other, and are held in their places by the pressure from without, they cannot give way. It seems to me that such a plan would work well in New Orleans.

A main sewer, in each of the streets running back from the river, at a depth of seven feet, would admit of lesser ones along the cross street, at a depth of five feet, into which should fall the surface gutters and the underground drains; which last should run at a depth of four feet into the centre of each square, under every store, alley way, &c. If all were placed at still a

5. "Levee street" is now Decatur Street, which begins at Canal Street and runs downriver, parallel to the Mississippi River, forming the riverside boundary of Jackson Square. The "(late) swamp" likely refers to the areas northwest of the Vieux Carré, now known as Broadmoor and Mid City, and the lowest part of the city. These areas remained swampy and undeveloped until the early twentieth century, when development of the Baldwin screw pump facilitated drainage and subsequent residential development.

greater depth it would be better. I find, from my own practice, that no land can be effectually drained, unless the drains are at a sufficient depth below the surface to leave at least one foot of soil unaffected by capillary attraction. Four feet suffices for this in open farm lands; but whether sufficient in a city where the stones of the pavement would very naturally assist in attracting moisture to the surface, I cannot say.

In my draining operations here, conducted on a small scale, and in the most economical manner—as all experiments should be—I have used, in place of tiles, the joints of our largest sized canes (the *Miegia Macrosperma*, of Michaud[6]) cutting them apart with a saw, so as to form tubes of one and a quarter to one inches long. These I have placed at a depth of three and a half feet, but will, in future, go as deep as four feet. The drains are thirty feet apart, and do not absolutely need a greater fall than one foot in a hundred. Two inch tubes, draining wet land, carefully laid, will keep themselves clear and be fully effective, with a fall of one-half what I have [illegible]. The tubes are carefully placed in the bottom of the drain, the end of the one fitting up to—not into—the end of the next. The water finds its way in at the joints, from below, they being rounded in above to prevent any entering from that direction, as it would then carry sediment with it. Although cane joints will last an age in such a position, yet I would prefer tiles, if to be had. By and bye I hope to have me a tile machine and make them.

But the thorough drainage of New Orleans should be made to yield a large revenue to the city, instead of being an expense. The sewerage waters of large towns, used for irrigating under-drained lands, have invariably, when thus applied, proved highly profitable. Witness the meadows at Edinburg and numberless others in Great Britain and on the Continent. Nor would the process, on drained lands, prove in the slightest degree injurious to health, which those not aware of the fact must take for granted at present; I have not leisure for explanation and proof.

Mr. Smith, of Deanston, in Scotland, the father, he may be called, of the systems of under-draining and subsoil-plowing, says in a recent report published by him: "To place the agricultural value of the at present wasted sewer water in another point of view, I have ascertained that the quantity of sewer-water due to a town of 50,000 inhabitants amounts to about 1,190,080,946

---

6. *Miegia macrosperma* or American cane, sometimes given as *Arundinaria gigantean subsp. Macrosperma (Michx) McClure.*

gallons per annum, which quantity will yield an annual application of 17,920 gallons to an extent of 66,410 acres.[7] Taking a general view of the subject, we may assume a clear revenue from the sewer-water of all towns, at £1 from each inhabitant." Suppose the sewer-water of New Orleans used for the irrigation of meadows of Bermuda grass (*Cynodon dactylon*), first thoroughly under-drained, on the vacant lands back of the city![8] The yield would be incalculable. Think of the abundant supply of rich milk from large dairies thus fed! In hay alone, at the price paid for that poor stuff, when compared with hay or Bermuda grass—Northern grown timothy hay—it would pay many pounds per acre for the use of the sewerage water. Those who know anything of irrigation will understand how it may be applied.

As the value of sewerage water for irrigation depends greatly upon its quantity as well as richness, the abundant supply from the Mississippi would give additional certainty to the experiment. The practical sufficiency of the fall stated by Dr. Barton may be estimated from the following table, by Professor Robinson. He found that—

A velocity at the bottom of a stream, in a second, of—

3 inches, will separate and lift up particles of fine clay.

6 inches, particles of fine sand.

8 inches, particles of coarse sand.

12 inches, will sweep along and lift up particles of fine gravel.

24 inches, particles of gravel one inch in diameter.

36 inches, angular stones of the size of an egg.

I would gladly extend these remarks, did time permit. But shall only add the assurance that with the exception of the additional expense of larger main drains, to be sufficient to carry off the sewerage water, the difficulties in the way of giving your city from four to six feet of dry soil, are no greater than are constantly being overcome in the draining [of] an equal extent of land for mere farming proposes.

Yours, respectfully,<br>THOMAS AFFLECK.

———◆———

7. Neither a geographic description nor an explanation of his calculations is given.

8. The reference to "sewer-water" here is to urban storm run-off and not to sanitary sewage, the system for which was not initiated in New Orleans until 1898.

## Summer Flowers for the Parlor—The Fuschia

Summer Flowers for the Parlor—The Fuschia
New Orleans *Daily Picayune,* May 9, 1851

*Eds. Picayune*—You will receive herewith a few specimens of that most beautiful parlor flower, most admired by the ladies of all Flora's gems, the *Fuschia*. As I knew you to be men of taste, and as your membership in the can't-get-away club deprives you in a great measure of the luxury of green fields, shady woods and bright flowers, I beg leave to contribute this, my mite, of the many good things we luxuriate in here in the country, that we may not seem given to a selfish enjoyment of them. As fruits ripen during their seasons you shall of these also partake. I would advise you, however, to *spell* [*sic*] each other in your daily labors for our amusement and instruction, and give us an opportunity—we of the country—of proving to you that we appreciate those labors. Be assured of your subscribers in the South the number who will not most warmly welcome any member of the *corps editorial* of their favorite paper is *very* few. Try it.

The Fuschia is native to South America, and numbers but a few *species;* but these, by the skill of florists, have been crossed in various ways, until the number of beautiful *varieties* has become very great. These form the most prominent objects of late years at all the horticultural exhibitions in Europe, no pains nor expense being spared in bringing them to the highest perfection of which the florist's skill is capable. You may form some idea of the dazzling beauty of what they consider *specimen plants,* by imagining one of these I send you towering to a height of ten to fifteen feet, the lower limbs spreading over and concealing the pot, each limb diminishing in length until the plant terminates in a single, swaying, delicate shoot, whilst from the base of every leaf-stalk hangs one or more of those lovely, gem-like brilliant flowers. My facilities for growing them do not admit of their being carried to a greater height than from four to six feet. The plants before you, if occasionally shifted into larger pots as those they are in become filled with roots, being careful that a few pieces of broken pot-shard are placed at the bottom of the pot, and using a soil for filling the added space composed of equal parts of rotted turf, leaf-mould and well decayed manure, watering once a week with a weak solution of guano or other liquid manure, and with rain water as often as needed, say once a day, taking care that the soil is not *kept* in a wet state, and once or twice a day sprinkling the entire plant with clear water, giving plenty of light and air, but not the full blaze of the noon-day sun, they will grow to a great size and continue to give a profusion of bloom all summer. This last they will do, however, as they now are, without any repotting. In the winter they will drop their leaves and cease to grow, when they may be set in any out-house till spring, where they can have light, air, and occasionally water.[9] In the spring they may be shaken clean of the soil, the branches cut back to three or four buds, and repotted or planted out of doors, to bloom in renewed bounty through another year.

The original species, the natives of the dense forests of South America, will not endure our summer's sun, but if scattered about a shaded lawn or garden the second year of their growth they form beautiful objects, supplying in some extent what our first gardens, lawns and shrubberies most need—ornamental undergrowth.

There are hundreds of ladies of taste detained through the summer in your city who would find great enjoyment and a relaxation from the weariness of passing our *glorious summers*—for glorious they are in the country—confined to close houses and sweltering streets, by growing a few fuschias, achimenes, glacinias, tree violets and other summer flowers in their parlors and halls, and on their porticos and balconies.[10] You remember how Charney, the prisoner of Fenestrella, let "captivity captive," in his care of his Picciola, seeing innumerable beauties in what to every other observer was but a paltry weed![11] Suppose his weed to have been a lovely Fuschia!

9. In this case, an "out-house" is likely a structure set apart from the main residence such as a shed or barn, and not a privy.

10. *Achimenes* is a genus of about twenty-five species of tropical and subtropical perennials, native to Mexico and Central America, with common names such as magic flowers, cupid's bower, widow's tears, and hot water plant. *Achimenes* species and hybrids are commonly grown as greenhouse plants, or outdoors as bedding plants in subtropical regions. Many of the species and their hybrids have large, brightly colored flowers. Generally easy to grow, they require rich well-drained soil, warmth, bright indirect light, constant moisture, and high humidity, similar to the conditions that fuchsias require. "Glacinias" may be the plant we know today as gloxinia, a colorful, tropical flowering plant native to the West Indies and Central and South America, belonging to the same family (*Gesneriaceae*) as *achimenes*. "Tree violets" are trees native to Australia, a curious plant to mention in association with the others; perhaps Affleck means another plant, the name of which is lost today.

11. *Picciola, or the Prisoner of Fenestrella; or, Captivity Captive,* by French novelist, poet, and dramatist Xavier Boniface Saintine (1798-1865), was a popular novel from 1836 in which the Comte de Charney, a political prisoner, maintains his sanity by the cultivation of a tiny flower growing between the paving stones of the prison yard. A nineteenth-century sentimental mas-

One of those I send you, marked "Anna," is a seedling of my own, and second to none of my imported plants.[12] I have a number of other beautiful seedlings, but none quite so fine as the one in question.

The neat balcony in front of the windows of your sanctum will suit these plants well, shading them when the sun shines too brightly on them

Yours, truly,

THOMAS AFFLECK.

P.S. By the way, what weather for a "May morning!" It is more like the last weeks of October. Not a drop of rain, at least not enough to lay the dust, for nearly a month; clear, bright and cold, with a brisk wind blowing. It is ruinous weather upon cotton and corn, the former especially, what little is yet standing or can be got up. And seed gets scarce from the general replanting. In my nursery the *undrained* and *unmanured* bud is almost burnt up.

---

terpiece and a classic of French literature, the novel was translated into many languages and earned the author international recognition. It is little known today.

12. We may conclude, by Affleck calling this seedling "my own," that it is named for his wife.

# 7

## On Roses

From ancient times to the present, no garden flower has attracted as much attention or has remained as popular as the rose. Joseph Beck, a prominent nineteenth-century nurseryman, wrote, "It is a flower beloved by everyone, not only in the present age, but has been in all ages past, and will no doubt continue to be the most prominent and desirable flower as long as the world stands. It may, with propriety, be styled the *Queen of flowers.*"[1]

Prior to the nineteenth century, roses were relatively simple varieties that had originated in Asia (China) or the Middle East (Persia). With increasing European interest in plants from exotic places, widespread cultivation and hybridization of roses began in the early nineteenth century. In *Nomenclature Raisonnée des Espèces, variétés et sous-variétés du Genre Rosier* (Paris: La Librairie de Madame Huzard, 1818), August de Pronville stated that there were fewer than two hundred varieties of roses in 1814; however, by midcentury there were over six thousand varieties because of keen public interest in the species and increased production through hybridization.

Much of this activity was centered in France, in part because of the attention of wealthy and aristocratic patrons such as the Empress Josephine (1763–1814), wife of Napoleon. A native of Martinique, Josephine had a general interest in the arts and natural sciences and a particular interest in botany and horticulture, and after her marriage to Napoleon in 1796 she acquired Malmaison, a property ten miles west of Paris. Starting in 1799, it grew from 640 acres to almost 1,800, with most of the site devoted to a garden designed in the fashionable late-eighteenth-century *jardin anglais* style, with winding paths, natural and apparently disorderly planted areas, exotic garden structures, and picturesque elements such as ruins, rock grottoes, and architectural fragments. Josephine used her influence as the wife of Napoleon to acquire plants from throughout the world. Plants and seeds

---

1. Joseph Beck, *The Flower-Garden; or, Breck's Book of Flowers* . . . (Boston: J. P. Jewitt and Company , 1856), 267.

regularly arrived from diplomats, scientists, military men, and governors of colonial outposts in efforts to advance their careers and standing with her husband. She became interested in roses in 1804 and set out to collect every known rose for her garden at Malmaison. Legend has it that she succeeded in collecting about 250 varieties, perhaps not a comprehensive collection but nonetheless an impressive one.

Josephine routinely corresponded with the leading botanists of the day in her quest for botanical knowledge, and she connected these men of science with important artists who then illustrated the scientists' works. One such contribution, both to science and to art, is the work of Pierre-Joseph Redouté (1759–1840), an unknown artist whom Josephine met in 1798. Her early patronage enabled Redouté to produce watercolor portfolios of the plants in Malmaison's gardens, notably *Les Liliaceés* (1802–16) and, most important, *Les Roses* (1817–24). Both works appeared in multi-volume editions and remain among the most highly regarded botanical illustrations ever produced. Even though *Les Roses* appeared after Josephine's death, its images instantly gained popular appeal and encouraged general enthusiasm for roses throughout Europe and America. Botanist Claude-Antoine Thory provided the text for *Les Roses,* and it was he who made the first attempt to create order in the chaos of rose genealogy. Since much of his work remains useful today and many of the 170 roses illustrated are still grown, *Les Roses* remains a standard reference on nineteenth-century roses.

Josephine's enthusiasm for roses and her celebrity as an arbiter of aristocratic taste encouraged an unprecedented international interest in rose culture. From this time onward, both amateur and professional horticulturists endeavored to develop new varieties to satisfy increasing public demand in Europe and America for new and different varieties. While roses are inevitably linked with Josephine and her garden at Malmaison, it was not until 1843, decades after Josephine's death, that the popular *Souvenir de Malmaison* rose appeared, commemorating both the garden and Josephine's passion for roses.

Among the most important nineteenth-century French rose cultivators were Jean Laffay (1795–1878), Jacques-Martin Cels (1740–1806), and Jean-Pierre Vibert (1777–1866), and the new varieties they introduced remain popular today. Laffay is generally credited with having developed hybrid perpetual roses, and he introduced many China and tea rose varieties into popular culture. Cels cultivated foreign plants for sale, many of which came from North America, contributing to the growing appetite in France for plants from exotic locations. As a young man, Vibert served in Napoleon's army, but following an injury, he returned to gardening near Malmaison, where he hybridized roses, fruit trees, and grapevines. Many of the roses mentioned in Affleck's articles were developed or introduced into commerce by these cultivators and their colleagues.

Roses were popular in colonial America and were found in gardens there from early colonial times. While many were passed along and transported from elsewhere to American soil, there were several commercial growers in the U.S. that supplied roses to North American consumers. Perhaps the most famous was William Prince's nursery, which began in 1790 in Flushing Landing, Long Island, New York, and passed through four successive generations of the Prince family.[2] A list from 1790 gives twelve roses, most of which have generic names ("Yellow rose," "white damask rose," "musk rose," "cinnamon rose," "thornless rose").[3] The Prince family had a significant impact on the growth of horticulture in the United States through its nursery and through the books that family members published, including William Prince's *A Short Treatise on Horticulture* (1828) and his son, William Robert Prince's, *Manual of Roses* (1846).

Another American grower of roses was John Champney of Charleston, South Carolina. In 1810 he grew from seed an accidental rose hybrid that he named *Champney's Pink Cluster*. It was sent to Louis Noisette, an eminent nurseryman in Paris, who renamed it *Blush Noisette,* and thus began a new class (the Noisettes) of roses.

We know from newspaper advertisements that roses were commercially available in New Orleans from the mid-1820s onward. In February 1825, the French Florist Gardeners

Have the honor of informing the public of their arrival in this city from Paris, with a beautiful collection of exotic plants, fruits trees of all kinds, shrubs, 150 varieties of the rose, [h]yacinths, daffodils, jonquils, tuberoses, amaryllis (very scarce), imperial crowns, a complete assortment

2. See http://riley.nal.usda.gov/nal_display/index.php?info_center=8&tax_level=4&tax_subject=158&topic_id=1982&level3_id=6419&level4_id=10847&level5_id=0&placement_default=0&test

3. This list is given in Ann Leighton, *American Gardens in the Eighteenth Century: "For Use or for Delight"* (Boston: Houghton Mifflin, 1976), 306–307.

of flower and kitchen vegetable seeds, to [sic] long to mention, and the catalogue of which, explaining the names and colours both in French and English, may be seen at their store, in Mr. Andry's house, Toulouse Street.

An advertisement from April 1844 notes that bookseller William Kern had received "Buist on the Culture of Roses!!!/The Rose Manual; containing accurate descriptions of all the finest varieties of Roses . . . with directions for their propagation."[4] And John M. Nelson's *Catalogue of Fruit, Shade and Ornamental Trees* . . . (1859) lists 160 roses available for sale at his Magnolia Nursery in New Orleans. Bernard M'Mahon's *The American Gardener's Calendar* (Philadelphia, 1806) lists thirty-nine varieties; by the eleventh edition of his work (1857), his calendar had evolved into a book about instructions for kitchen, fruit and flower gardens, orchards, vineyards, nurseries, and greenhouses.[5]

Among the first books published in America on roses was Robert Buist's *The Rose Manual* (Philadelphia: Carey and Hart, 1844). A Scotsman, Buist immigrated to the United States and settled in Philadelphia in 1828. His *Rose Manual* was written for the American audience, most of which was female, and it took two years to write. It was widely circulated and had appeared in four editions by 1854. At this time, Buist had the most extensive collection of roses in America, and this work is based on his experiences as well as on those of others in the region. Buist's *Manual*, listing nearly four hundred varieties of roses by name, is written with great attention to detail and is a complete discussion of rose culture.

William Robert Prince (1795–1869) was the fourth owner of his family's Linnaean Botanic Garden and Nursery in Flushing, New York, and his *Manual of Roses*, according to horticultural historian U. P. Hedrick, is a "much better work, but never got beyond the first edition, probably because Buist wrote in a simpler style."[6] Other books of the period that discuss roses are Louisa Johnson's *Every Lady Her Own Flower Gardener* (New Haven: S. Babcock, 1844); Samuel Bowne Parsons,[7] *The Rose: Its History, Poetry, Culture,*

4. Robert Buist was an influential American horticulturist. Discussion follows.

5. For M'Mahon's 1806 list, see Ann Leighton, *American Gardens of the Nineteenth Century: "For Comfort and Affluence"* (Amherst: University of Massachusetts Press, 1987), 346–48.

6. Y. P. Hedrick, *A History of Horticulture in America to 1860* (New York: Oxford University Press, 1950), 482.

7. Samuel B. Parsons (1819-1906) founded a famous nursery in Flushing, New York, in 1838; his son, Samuel Bowne Parsons Jr. (1844-1923), became a landscape architect and worked with

*and Classification* (New York: Wiley & Putnam, 1847); *The American Rose Culturist* (written anonymously; New York: C. M Saxton, 1852); John T. C. Clark's *The Amateur's Guide and Flower-Garden Directory* (Washington, D.C.: Taylor and Maury, 1856); and Joseph Breck's *The Flower-Garden* (Boston: John P. Jewett and Co., 1856).[8]

Shorter works appeared regularly, however, in nineteenth-century periodicals and agricultural papers of the period, and often they are contributions from subscribers or regular contributors, as is the case with the eight articles by Affleck given here. When considered within the broader context of geography and available resources, we see that Affleck's articles on roses, published in newspaper format rather than in book form, did not have as wide an impact as other published resources available in this period, and doubtless that explains the absence of Affleck's articles in histories of rose-growing in America. Nevertheless, these articles are of value because they address local growing situations and document which roses were available locally, even if assigning modern names to Affleck's lists is often problematic.

Writing in the fall of 1856, from his Southern Nurseries in Adams County (just twelve years after Buist's *Rose Manual* and ten years after Prince's *Manual of Roses* appeared), Affleck produced a discussion of roses that appeared over a two-month series in the New Orleans *Daily Picayune*. For years, Affleck had supplied horticultural observations and advice to numerous newspapers and had annually published his *Southern Rural Almanac* in New Orleans since the late 1840s. His lengthy comments on roses—over 8,500 words—were written, according to their bylines, in October and November of 1856 from his nursery in Washington. Perhaps Affleck chose the *Daily Picayune* as the vehicle for his observations because lengthy articles such as these would not easily fit into his almanac's format. Or perhaps he determined that publishing these articles in a New Orleans newspaper was a good

Calvert Vaux; they were partners from 1887 to 1895. Parsons Jr., a founder of the American Society of Landscape Architects (1899) and later its president (1905-1907), is best known today for his work as superintendent of public parks in New York City. His book *The Art of Landscape Architecture* (New York: G. P. Putnam's Sons, 1915), once used extensively as a textbook, is today largely forgotten.

8. For early American books on roses and rose culture in America, see Hedrick, *A History of Horticulture in America to 1860,* 270–73.

way to reinforce his standing with local clients (of whom he had many) and to reach new customers for his nursery. Curiously, he mentions his *Almanac* only once (and in passing, in the eighth installment) and does not discuss which plants his nursery had in stock. However, he mentions his knowledge of these roses comes from personal experience, and therefore the numerous examples given in his articles are roses suited to the growing conditions of this region. In his *Almanac,* Affleck gives plants available from his Southern Nurseries; an advertisement from 1854 shows numerous roses available, most of which are mentioned in these articles.

The roses that Affleck discusses are primarily roses developed in France, but one exception is the "very curious" Green Rose, introduced by American John Smith in 1827, and, Affleck notes, "[A]ccording to the newspapers this rose seems to command enormous prices in Europe." As an importer of plants from growers in Paris, Affleck likely had access to most new varieties available from European sources. In listing the names, he often drops French spellings and punctuation, and sometimes names are given in English translation (original French names, French punctuation, and dates of introduction, when known, are given as editor's notes). He describes "La Quintinie," for instance, as being "a new and superb rose"; it was introduced in 1853 by Desiré Thomas, who was actively breeding between 1846 and 1863. Comparing Affleck's descriptions with those in modern sources usually verifies that modern names match nineteenth-century names; slight differences appear, however, usually in the highly subjective descriptions of color (Affleck's "delicate rose" might be today's "pale pink").

As elsewhere, minor corrections in spelling and punctuation have been silently made for ease of reading. In general, Affleck set the names of roses in italics, and that form has been preserved here. Notes have been added to supplement Affleck's text or to expand on information given. Matching the roses listed in Affleck's articles against three reputable electronic sources for "old roses" reveals that most of the roses he lists were of French origin and date from the late eighteenth century to the mid-1850s.[9]

One of America's largest suppliers of "antique" roses, and one of the best sources of information about them, is the Antique Rose Emporium of Brenham, Texas, coincidentally the rural community in East Texas where

---

9. http://www.antiqueroseemporium.com/; http://davesgarden.com/guides/pf/finder/index .php?sname=Roses; and http://www.helpmefind.com/rose/index.php

Affleck settled upon leaving Mississippi in the late 1850s. Located just a few miles from the site of Affleck's home and his Central Nurseries, the Antique Rose Emporium began in 1982 as a source of old roses. Much of its stock has come from cuttings taken (as they say, "rustled") from abandoned cemeteries, homesteads, and derelict properties throughout Texas and surrounding states. Some of the roses offered today are those Affleck discusses in the following articles; perhaps some may even have come from plants purchased by Affleck's clients in the region. In 1989 the Antique Rose Emporium introduced a rose named for Thomas Affleck.

—Ed.

## The Rose

November 5, 1856
The Rose.

The Southern Nurseries
Washington, Adams County, Miss., Oct. 27.

It is rare that a "mail day" passes without bringing one or more letters inquiring "what shall I do to my rose bushes? How am I to prune them? Must they be manured, and how, and with what? And what varieties do you consider the most beautiful and desirable?" and so on.

Now, as these inquiries come almost invariably from the ladies, and as it is impossible for me to reply, at any length, to each and all, and as we have no treatise on the subject at all adapted to the South, entering as fully into particulars as the ladies seem to desire, I must appeal to you, Messrs. Editors, to help me out of the difficulty.[10] Let the types multiply, and scatter to the extent of your circulation, the following reply to one of the fair inquirers, and you will gratify your lay readers who have a taste for the Queen of Flowers—and who of them has not?—and at the same time relieve me from a quandary!

Although, during the last ten years, it would seem that the perfection of beauty had been attained, in the lovely roses produced and disseminated during that time, we have every now and then something new and striking brought out. And yet, there are many which have been cultivated, prized and admired for years that are altogether unsurpassed.

---

10. It was not until 1912 that a book appeared for rose enthusiasts of the American South, *Everblooming Roses,* written by Mississippi native Georgia Torrey Drennan. Born in 1843, Drennan was the daughter of a wealthy plantation owner in Holmes County, Mississippi. Married in 1861 at age seventeen, she was the young wife of a lieutenant serving in the Confederacy, and his letter to her describing the Battle of Vicksburg in 1863 is an important description of the siege of that community. Her book, subtitled *For the Out-door Garden of the Amateur,* makes no reference to where she gained her experience in growing roses. Perhaps Drennan, who would have been in her late sixties when her rose book appeared, did not want to call attention to the fact that she was from the South by discussing her heritage or targeting her book to a regional audience. Though not explicitly aimed at audiences in the South, the book was highly regarded when published but has over time sunk into obscurity. Drennan discusses numerous rose varieties and gives many examples of "everblooming roses." Her personal comments on each rose are particularly valuable.

Before, however, describing individual varieties, or speaking of their treatment, it may be well to explain that there are two great classes of roses— those which bloom only once a year, in the spring or early summer, and hence known as SUMMER roses; and those which, in addition to the abundant crop of flowers which they give at that season, bloom again after almost every shower that falls; or only in the autumn, after the usual rains of that period, and hence are all called AUTUMNAL.

This is rather a forced classification, however; but as it has been generally adopted by European growers, it is also very generally retained here.

So much of the attention of rose growers has been directed, of late years, to the perpetual and autumnal bloomers, and so great a variety produced, in habit of growth, color, fragrance, form of flower, &c, that many of the summer-flowering ones, which were, until recently, deemed indispensable, are now supplanted. There are still, however, some of these retained as beautiful favorites which none can bring themselves to discard for the sole and only reason that, like the true woman, they are somewhat chary of a too free exhibition of their charms!

Who, for example, will reject the *Mosses*, the *Banksias*, the brilliant and variegated *Provence* and *French Roses*, the *Sweet Briars*, the graceful *Ayrshires*, pure and bright yellows and copper colored *Austrian Briars* and *Double Yellows*, or the hardy and showy, rampant *Double Prairie Roses*? Yet all of these, except some few new *Mosses*, bloom but once a year—early in the spring and summer. Then, however, they give us gems of the purest water, and in great abundance; the blooms continuing in perfection for a great length of time.

The BANKSIAS, or LADY BANKS ROSES, from one of the most delicately beautiful groups we have.[11] Being too delicate for the climate of the Northern States or of any but the most Southern parts of England, they are not spoken of as they deserve, in the only treatises we have; with us, south of Tennessee and North Carolina, they are perfectly hardy; and from their graceful weeping habit, vigorous growth and rich evergreen foliage, they are at all times highly ornamental; and when in bloom are surpassingly beautiful.

11. Lady Banks rose (*Rosa banksiae*), a native of China, is named for the wife of eminent English botanist Sir Joseph Banks (1743–1820), who sent William Kerr (d. 1814) on a plant-collecting expedition to China in 1804. Kerr remained in the Orient for four years and sent back numerous plants now commonly cultivated, among them the Lady Banks rose.

The *double white* and *double yellow* are the only varieties we have had amongst us, until quite lately; unless, indeed, we include the *Cherokee hedging rose,* which belongs to the same group.[12] The *double white,* in addition to the beauty of its pretty clusters of small very double flowers, produced in vast abundance, has also a very delightful and delicate violet perfume to recommend it still further. The *double yellow* has no fragrance but produces its pretty wreaths of small double buff-colored flowers in equal profusion. There are in this country plants of these two roses of vast size: some having stems of ten inches to a foot in diameter, and embracing a vast circle in the sweep of their branches.

I have some strong plants of some newly introduced varieties of the Banksias, which will bloom most profusely this coming spring. Last winter's cold was so severe as to injure the young shoots so far as to prevent their flowering freely—with one exception.

*Alba grandiflora,* the *single white* and *spinosa lutea* are said to be all three very beautiful.[13] If equally so with the *Fortuniana,* I shall be well pleased. This was in perfect bloom, notwithstanding the cold, and proved to be as fragrant as the white, and fully as profuse a flower, each bloom in the thick clusters being larger when expanded than a half dollar, and of the purest white.

*Fortune's Chinese Yellow,* though not properly a Banksia, is yet nearer akin to this than to any other group. This is a decided novelty, a vigorous runner, with rich, glossy foliage, and an infinite profusion of flowers of a beautiful golden yellow color, with a delicate penciling of rose color on the outside of each petal. It is fully semi-double. The color is so peculiar as to attract the eye at a great distance.

All of this class require rich, deep soil, and an annual top-dressing of rough manure in December. For pruning, the shoots should be thinned out during winter; but those shoots that are left must not be shortened or otherwise cut until after they have done flowering; nor it is necessary at all, un-

12. Affleck advocated using the native Cherokee rose (*Rosa laevigata*) as a barrier, hedgerow, and fence in the South, where it grew in abundance. Frederick Law Olmsted commented on it in the Felicianas, north of Baton Rouge: "The roadside fences are generally hedges of roses—Cherokee and sweet brier." *A Journey in the Back Country in the Winter of 1853–54* (New York: Mason Bros., 1860) 13–14.

13. Today the most common Lady Banks rose is the double yellow version, *Rosa banksiae Lutea.*

less for the purpose of training to cover some particular space. They may be nailed to the walls of a house, to cover or conceal what may be otherwise unsightly, or arbors may be formed of them; or, if planted in the lawn or open grounds, all of them may be made to assume the forms of graceful weeping trees, by training to a stout cedar or locust post.

[To be continued.]

————◆————

### The Rose—№ 2

November 9:
The Rose—№ 2
[Continued.]

The Southern Nurseries,
Washington, Adams Co., Miss.

The class known as Moss roses have always been and ever will be general favorites. The half opened bud, so delicately and modestly concealed in its mossy covering, is one of the most lovely of floral gems. The complaint is often made that this class does not thrive and bloom well in the South. But if such is the case, it is the fault of the kind or of the cultivator. Some of them are weak in their habit of growth, and do not grow well unless grafted or budded upon the *Rosa Manettii*[14]—a strong-growing, most vigorous variety I employ for that purpose. And all, even those of strong growth, require deep, rich soil and liberal winter manuring to bloom well. Like many others of our finest roses, as *Devoniensis, Chromatella, Augusta,* &c., they give us much finer and more perfect blooms, and in greater profusion, when two or three years old, and well grown, than when younger.[15] This past spring (1856) gave me such a display of moss roses as, I suspect, was never before seen in the south, five hundred half opened buds and fully expanded flowers were repeatedly counted of a morning on single plants.

All of the following grow, thrive and bloom here in great perfection:

*Blush*—As its name implies, of a delicate blush color; an old but fine and distinct variety.

---

14. *Rosa chinensis manettii,* a noisette from 1835.
15. *Devoniensis* is a tea rose from 1838; *Chromatella* is a noisette from 1843.

*Bourbon*—An astonishingly strong grower; makes quite a tree. The foliage is very large, abundant, and richly glossy like the *Bourbon* roses; the buds quite mossy, opening well; color bright rose.

*Celina*[16]—By far the finest and brightest of the crimson-colored moss roses; with leaves of a dark glossy green; the moss, too, enveloping the buds and stems, is of the same dark color. The habit of growth is trailing. I, therefore, intend working a few strong stems of the Manettii at a height of two or three feet, to form trees. If successful, the effect will be very pretty.

*Comtesse de Murinais*[17]—Another very vigorous grower; equally beautiful in the bud, and, when expanded, which is not often the case with this class, color pale flesh, changing to white, and very pretty.

*Crimson* or *Tinwell*[18]—a well-known old rose. Buds, leaves and branches all more or less mossy. Color, a beautiful shade of crimson. Habit moderately strong.

*Delatante*—Bloomed for the first time with me this past spring. Color, bright glossy rose; buds beautiful; flowers large; habit vigorous and robust.

*Hooker's Blush*—Very pretty, delicate and quite mossy. The ladies rarely permit a bud to expand.

*Laneii*[19]—Color deep rose; in some seasons rosy crimson tinted with purple; buds richly mossy; a vigorous and fine rose.

*Luxembourg*—A fine old rose; well known; color crimson, and very pretty.

*Perpetual White*—Pure white; blooms in large clusters: a perfect *remontant,* blooming out afresh after every summer shower. The leaves are fragrant. Like those of the sweetbrier.

*Princesse Adelaide*—A very strong-growing rose; equally so with *Bourbon;* with large, rich, dark foliage; producing abundance of mossy buds, expanding well; color a lively pink.

*Prolific*—Rose-colored; exactly like the *old moss,* but dwarfer in habit; and a more profuse bloomer.

*White Bath*—Pure white; a most beautiful rose; with lovely mossy buds. Requires a rich soil to bloom in perfection.

16. Introduced 1855.
17. Introduced 1843.
18. Pre-1846; also known as the Tinwell Moss rose.
19. Lane's Moss; from 1846.

All of this family or class should be pruned after the season of flowering, to encourage a strong growth of new wood upon which it is that the next spring's blooms are produced. After that they need only a little thinning out and regulating. The following winter, those strong shoots which are out of place may, however, be shortened or removed, and weak shoots also shortened. When it is desired to form large trees, those buds which show themselves when no branches are wanted, must be rubbed off, and especially those which may push from near the ground. I again repeat that no class of roses thrives more perfectly in the south than this. They all require good deep soil, and heavy surface manuring in the winter.

The FRENCH and PROVENCE roses are now chiefly prized for their highly variegated and striped varieties. It has been the anxious aim of Rose-growers to produce amongst the *Perpetuals* some which might equal these; but hitherto in vain. I had a neighbor who was induced by the beautifully colored drawing exhibited by a French peddler of plants and flowers, who guiled the citizens of Natchez, New Orleans and others of our cities in the Southwest for several years past, to pay some $5 or more for a plant of "THE *carnation-striped perpetual moss rose!*" It bloomed in my hands and proved to be a very indifferent rose-colored hybrid-perpetual! Had it equaled the drawing it would have been cheap at $100.

*Œillet parfait* is one of the best of the *Gallica* or French roses. It is white, finely striped with rose and bright red; but with time has, as yet, hardly equaled its European reputation.

*Eulalie le Brun,* on the contrary far surpasses my expectations, and is, indeed, most delicately beautiful. It has a ground color of pale pink, clearly and prettily striped with white.

*Perle des Panachées*—White, striped with red and purple; a pretty rose, though of rather delicate habit.

*Habit d'Orleans*—Red, with yellowish white stripes; but though singular and pretty, is only semi-double.

*Madame Zoutman*—A damask rose, with all the fragrance of that class, is a pretty, delicate cream color, tinted with fawn.

*Snowball* is another of the same class: pure white and very pretty. These two are improvements on what some often asked for, "the old fashioned white, fragrant rose."

*Jenny* is a hybrid-China rose; hardy, a most profuse bloomer, flowers of the most perfect form and of a deep rosy lilac.

*La Fontaine,* another of the same class, of a brilliant crimson color, and like Jenny, a most robust grower.

*Descartes,* yet another, of a singular purplish rose-color, of fine form and a great bloomer.

*George IV,* also a hybrid China, is the darkest rose in cultivation—a crimson of so dark a shade as to be usually called the "black rose." It is a hardy, vigorous, beautiful, free-blooming variety, so distinct in color as to be very desirable.

*Coupe d'Hébé* [20] is a hybrid-Bourbon, and in my estimation the most perfectly formed rose we have. Its color is a delicate, yet warm, bright rose, and very beautiful.

*Great Western,* is another of the same class: a most robust rose, producing flowers of the very largest size, and of a deep reddish crimson, very showy and attractive.

All of these require treatment as directed for the mosses.

THOS. AFFLECK.

[To be continued.]

---

### The Rose—№ 3

November 14:
The Rose—№ 3
[Continued.]

Southern Nurseries,
Washington, Miss., October 1856.

The SWEET-BRIARS are general favorites. Nearly every one has pleasant associations connected with them. The delicate fragrance of the leaves has, hitherto, been all that rendered them particularly desirable. But, of late years, new varieties have been produced which retain all the pleasant fragrance of the foliage, and at same time yield brilliant double flowers.

The AYRSHIRES from a class of extremely vigorous runners, very hardy, blooming early in the season, and withal very pretty. They are employed to cover rocky, rough or otherwise unsightly spots, or treated as I suggested for

20. Introduced 1840.

the Banksias, they form equally pretty objects on the lawn. Or, still better if budded upon tall stems of the *Manettii,* where they grow vigorously, making the prettiest of all weeping trees. Of the Ayrshires, two of the prettiest are—

*Dundee Rambler,*[21] whose shoots are very long and flexible, and are covered in the spring with pretty little roses, white often edged with pink.

*Splendens*[22] has the same habit of growth, and finely cupped creamy-blush flowers, produced in great plenty, and are in the bud especially beautiful. They have a peculiar and agreeable "annis-scented" fragrance.

The AUSTRIAN BRIARS[23] form a very distinct class. Their stems are reddish colored, and literally covered with softish spines of the same color. They bear bright yellow or copper colored flowers; and have all more or less of a running habit.

*Harrisonii*[24] is perhaps the most beautiful purely yellow rose we have, and is very pretty indeed when well grown. It is said to have originated in America, from seeds of one of the Austrian briars.

*William's Double Yellow* is also bright yellow, pure and unadulterated with any other tinge, dwarfish in its habit, and very pretty.

The PRAIRIE ROSES were produced near Baltimore, from seeds of the native roses of the Northwestern prairies.[25] They are quite double and very showy, vigorous, rampant runners, and perfectly hardy. When planted in the lawn, and trained to a stout post, in rich soil, they can scarcely be surpassed as showy, attractive ornaments. The following are desirable:

*Baltimore Belle*—pure blush, a great bloomer, and very double.

*Purpurea*—equally showy, flowers purplish, and very double.

*Queen of the Prairies*—rosy, red, faintly striped with white.

*Superba* or *Pallida*—pale blush and very pretty.

I have devoted considerable space to the SUMMER ROSES, but no more than that they are well entitled to; for although where there is space for only a few, the more perpetual bloomers are certainly the most desirable; yet the

21. Introduced prior to 1837.

22. Introduced prior to 1837.

23. *Rosa lutea.*

24. *Rosa foetida harisonii* or *R. Harisonii;* bred by American George Folliott Harisson, ca. 1824.

25. Baltimore nurseryman Samuel Feast (1796–1868) propagated roses using the native *R.rubifolia* (now *R. Setigara*) to develop the Prairie roses given here, all of which date from around 1843. Feast was one of the founders of the Maryland Horticultural Society in 1830.

variety of habit, the brilliancy of color, and the diversity of ornamental purposes which these other are susceptible, render them indispensable whenever the garden and grounds are of any but the most moderate extent.

All of this class are hardy, suffering more from a long summer's drought than from the severest cold. Like all roses, they require good soil to grow and bloom well but will generally, with the exception of the *Mosses and Banksias* do well where the Perpetuals would never yield a flower worth of notice.

The AUTUMNAL ROSES are divided into several classes or groups.

HYBRID-PERPETUALS, which bloom freely through all of the growing weather of spring and early summer, and begin again on the setting in of the fall rains, when many of them produce flowers much finer than those they bore in the spring. Some of them bloom after every summer shower. They are generally highly fragrant. Their stems and branches are nearly all quite thorny and their leaves are rough.

The DAMASK PERPETUALS have nothing very decided to distinguish them from the last-named. They have more of the old Damask Rose fragrance and habit of growth, perhaps; have erect, still and very thorny shoots, and bloom very freely.

The BOURBONS are the real *perpetuals* in their habit of blooming—continuing in flower, more or less perfectly, from frost to frost again. They are less thorny than either of the last named; and the thorns are strong, and usually light colored. The stems and branches are generally stiff, smooth, glossy, and add much to the beauty of the plant. The petals or flower leaves are also usually thick, and have a peculiar satin-like or burnished appearance, that gives them additional attractiveness, and enables them to endure the heat of a summer's sun hot enough to cause all others to flag. Their growth is luxuriant, and they are quite hardy.

The CHINA ROSE forms another class of very constant flowers; even more so than the BOURBON, and equally so with the NOISETTES and TEAS, which are, in fact, varieties of the CHINA.

The HYBRID-PERPETUALS comprise so very large and beautiful a group, that it is an extremely difficult matter to make a selection of the most desirable. I tried to make a selection of ten, but soon extended it to twenty; and before I got through, found I had described double that number; not one of which I should like to reject.

*Augustine Mouchelet* is a deep and richly shaded crimson rose, very fragrant, and a moderately vigorous grower.

*August Mie* is a new and very beautiful color, a light glossy pink, and flowers of the most perfect form.

*Baronne Prevost* I consider one of the very best of the class; a very large and perfectly formed and fragrant rose of a brilliant rose color, a constant bloomer and vigorous grower. Wherever there is room, a large bed of it should be made. Makes a magnificent pillar, and is desirable in every collection.

*Blanche (Vibert)* is a pretty, fragrant pure white rose, blooming in clusters, which are prettily nestled in leaves. In the fall the blooms are of a pale straw color in the centre.

*Clementine Dural* I have only had one season; but promises to maintain its European reputation. It is a delicate rosy lilac, and of good form.

*Comte d'Eu* is of very dwarfish growth, and is excellent for grouping; color, brilliant carmine, inclining to scarlet, and very pretty and attractive.

THOMAS AFFLECK<br>[To be continued.]

———◆———

### The Rose—№ 4

November 18:
The Rose—№ 4
[Continued.]

The Southern Nurseries.
Washington, Miss., Oct. 1856.

In my last I began to describe some of the most beautiful of the HYBRID-PERPETUALS.

*Comte Ory*—is new, pretty, and distinct; color, bright carmine, beautifully shaded with violet.

*Comte de Paris*—is one of the most desirable; color, bright crimson, with a shade of lilac; of very perfect form, double and fragrant, growing and blowing all summer.

*Duchesse d'Orleans*—bright rose of a lively shade; large, finely shaped and excellent.

*Duchesse de Montpensier*—shaded black, shape perfect; a distinct and very striking rose.

*Duchess of Sutherland*—a superb old variety, of a bright rose color, beautifully cupped; growth and foliage both luxuriant.

*Earl Talbot*—one of the largest and most *cabbaged* roses we have, very double and fragrant; color, deep rose; very desirable.

*Edward Jesse*—light carmine, with a fine violet tinge; large, fragrant and very showy.

*Genie de Chateaubriand* [i.e., *Génie de Châteaubriand*]—a distinct and showy variety, of a singularly beautiful violet red, shaded; the underside of the petals of a whitish shade; very large, and a constant bloomer; altogether desirable.

*Géant des Batailles*—"Giant of Battles"—well named; one of the most beautiful of roses; in habit of growth, brilliancy of color and perfection of form, altogether distinct; color, glowing crimson, approaching to scarlet.

*General Merlin* [*Général Merlin*]—Lovely rose-color, with flowers finely shaped; quite attractive.

*Islande d'Arragon*—Light rose; an old and fine variety; blooms freely.

*Lady Alide Peel*—Deep pink, often veined with red; shape perfect; a charming variety.

*La Bedoyene*—A distinct and very beautiful variety; dark but most brilliant carmine inclining to scarlet; a great bloomer.

*La Reine* (Laffay's) is one of the largest, most perfect and finest of roses when well grown; color brilliant, glossy rose, and very fragrant.

*Louis Bonaparte*—A fine fragrant, constant blooming variety, of a rosy crimson color.

*Louis Paget*—is a distinct and showy, free-blooming rose; color a singularly delicate pink.

*Madame Aimée*—Somewhat of the style of Marquise Bocella; pale flesh and beautifully delicate.

*Madame Campbell d'Islay*—A robust, vigorous grower; color veined rose; a seedling of *La Reine,* which generally gives flowers curiously striped with red, and is very pretty. This has been and is still sold at a high price as a "new, striped perpetual rose."

*Madame Laffay*—Rosy crimson; one of the most excellent roses in this group; indispensable.

*Madame Rivers*—Pale flesh, nearly white, perfectly shaped, and very distinct and beautiful.

*Marquise Bocella* is one of my most especial favorites; nearly always in bloom and always lovely. Habit, dwarfish, still, effect and very distinct; flowers large, very double and fragrant; of a pale rose color, at times a mere blush, and at others, and especially in autumn, singularly and prettily variegated.

*Marquis of Ailsa*—A brilliant crimson rose, of great beauty; very fragrant; habit robust and vigorous.

*Mrs. Eliott*—Light crimson, tinted with lilac; a fine, fragrant old rose, of robust habit.

*Paul Dupuy*[26]—Is a new and very desirable rose, of perfect shape; color, crimson, beautifully tinted with violet; altogether beautiful.

*Pius IX*[27]—Is another superb rose; a rampant grower, and equally free and constant bloomer; makes a superb pillar rose. The color is most singular deep purplish rose, with lighter shading towards the centre, differing in tinge with the season; shape of the bloom very perfect.

*Pompon*—Rosy pink; flowers very small, gem-like and sweet; forms a charming little bush.

*Prince Albert*—One of the beautiful and desirable of this class; flowers double, large, finely formed and unusually fragrant; color, deep crimson and purple; almost always in bloom.

*Robin Hood*—Rosy carmine; habit very graceful: blooming in large clusters; a charming rose.

*Stanwell's Perpetual Scotch Rose*—Has the delicate, trailing habit of that family, but is a true perpetual bloomer. Flowers of fine form; in color pale blush and very fragrant; when budded on a stem of the Manettii, say three feet high, it becomes a pretty weeping bush.

*Souvenir de Leveson Gover*[28]—A handsome and showy new rose; color, bright shaded red; large and beautiful.

*Wm. Jessé*—Certainly one of the most beautiful very large roses. Large and superb, opening freely; color, crimson, with a rich tinge of lilac.

There are new varieties constantly being produced. But the fact that they are new amounts to nothing, if they are not, at [the] same time, distinct and beautiful; and add something in habit, color, form, &c, that may be really desirable, to those we already have.

26. Introduced by François-Narcisse Dupuy-Jamain in 1851.

27. Introduced by Jean-Pierre Vibert, 1849.

28. Also given as *Souvenir de Leweson Gower* and *Souvenir de Leveson-Gower;* introduced by Guillot in 1852.

I have had a large number now under trial, in addition to those enumerated. It requires more than one season, however, to prove a new rose, and determine whether it is suited to the climate or not. Many of the very finest do not produce really fine blooms until the plants have attained a season or two's growth and become completely established; and none of them bloom well unless in deep, rich soil, and annually manured and tended.

It is difficult to describe the color of many of these roses in words. For instance, "crimson tinted with lilac" may be employed to describe, and that truly, the color of two roses; which are, however, really unlike each other in color. And no words can describe the intense brilliancy of color of Giant of Battles, or the singular beauty of Pius IX, or the clear warmth of color of Marquise Bocella, or the delicate stripes and veins in the petals of M'me Campbell d'Islay.

Then there are roses which bloom in the highest perfection in the spring and summer but in the fall, lack clearness of color, perfect form, &c.; whilst another nearly resembling it blooms in the fall in the highest perfection, but in the spring is not particularly attractive. And for this, it is, that a very considerable variety is required in order to have roses in perfection at all seasons.

THOMAS AFFLECK.

[To be continued.]

———◆———

**The Rose—№ 5**

November 27:
The Rose—№ 5.
[Continued.]

The Southern Nurseries,
Washington, Adams County, Miss., October, 1856.

The class of DAMASK PERPETUALS includes some of the most perfect, fragrant and constant blooming roses we have. Therefore several of those classes as Hybrid Perpetuals which might, with much more propriety, be placed here: such as Blanc (Vibert),[29] Madam Aimée, Marquise Bocella, &c,

29. Blanc de Vibert, developed by French rose breeder Jean-Pierre Vibert.

*Bernard* is an extremely beautiful and fragrant rose, of dwarf habit, and a very frequent bloomer. The color may be called salmon—pink, and has a warm glow about it that is very pleasing.

*Du Roi* or *Crimson Perpetual*—This old rose is yet unequaled. The fragrance is so delightful, the crimson color so rich, the form so perfect, and the habit of blooming so constant, that it is unsurpassed. Whenever the grounds admit of it, beds or groups of this rose should be made.

*Duchess de Rohan*—A large and fine rose, rich and fragrant; color red, shaded.

*Madame Thelier*[30] is a pretty rose; color, a delicate and very pretty pink; habit, slender.

*Mogador,* or *Superb Crimson*—brilliant crimson, a fine and distinct variety; by many esteemed above Du Roi, from which it varies greatly in shade of colors.

All of the DAMASK and HYBRID-PERPETUALS require and *must have* a superabundant supply of food to enable them to bloom well. It is, therefore, folly to plant them in dry and untilled lawns or door-yards, suffering the grass to grow close up to their stems, and leaving them unmanured for years. Under these circumstances, even those which usually make the finest display of autumn flowers will scarcely ever push forth a second series.

The spot selected for a Rosarium should, when at all practicable, be one where the soil is deep, rich, and not too sandy. If not deep and rich, it must be made so by trenching and thorough manuring. Even sandy, dry soil may be rendered less objectionable by being well manured with cow droppings, and by liberal surface dressings of manure late in the fall or winter. I have found that a double handful of bones, either ground to powder, or broken in pieces as small as can be conveniently done by an old axe, and well mixed with the soil filled in around the rose, to prove quite beneficial. However, the best animal application that can be made to roses in the way of surface manuring, is *human ordure,* made thin like a thick gruel, and poured into a shallow, wide basin, surrounding each plant, say in December; and then covered with enough of earth to conceal anything offensive. The rains of winter carry down the strength of the manure, to the vast improvement of growth and bloom.

The pruning required by these classes consists in cutting out with a sharp knife (shears may answer, if such as make a smooth cut without bruising the

30. Also *Madame Tellier* or *Madame Thélier.*

bark be used, though the knife is best) all of the weak and straggling shoots smoothly and close to the stem, and in shortening the stronger shoots in proportion to their strength, cutting the weaker ones back to within four or six buds, and leaving the strong ones a foot long or more. This should be done in December, if all growth has been checked. When the spring blooming is over, again cut out slender shoots, leaving half of the others their full length, and cutting the remainder to half their length. The shoots thus cut back will probably bloom after abundant rains in June, or if thoroughly drenched with water; if not, they will force strong shoots which will bloom freely in the fall.

The class of BOURBON roses I have already spoken of as the best adapted of all to the extreme South, inasmuch as the flowers stand uninjured by the hottest summer sun. It embraces, too, some of the most beautiful of known roses.

*Chaillot*—A very pretty variety, with large clusters of delicate rose-colored blossoms.

*Enfant d'Ajaccio* offers a sort of connecting link between the Noisettes and Bourbons: a fragrant, attractive, showy rose, of a brilliant scarlet color, shaded with crimson; suited to form a pillar.

*Gerbe de Rose,* too, can hardly be called a pure Bourbon. Is of very strong habit, with rich foliage; color bright rose, edged and shaded with white.

*Gloire de France,*[31] or *Monthly Cabbage Rose,* is a fine old variety, and an abundant bloomer; color light rose, and very pretty.

*Gloire de la Guillotiere* [Guillotière]—light rose, large and superb, and a free bloomer.

*Hermosa*—is an old variety; still one of the best, and is well named, "the beautiful." Color, delicate rose; very double, and perfect, and nearly always in bloom.

*La Quintinie*—a new and superb rose, of a deep crimson, purple color; beautiful, large and finely shaped; habit slender.

*Leveson Gower*—of a deep and very singular rose color; brilliant and quite distinct; blooms very large and double, and equal in form to Souvenir de la Malmaison.

---

31. This likely is the rose introduced in France by Neumann in 1824. There are other roses with this name, including a hybrid centifolia gallica (Bizard, 1828), a gallica (Hardy, introduced pre-1836), and a hybrid perpetual (Margottin Père & Fils, 1852).

*Madame Desprez*—a robust rose, blooming almost constantly in immense clusters of cupped and very double rosy-lilac flowers of the greatest beauty.

*Madame Nerard* [Nérard]—a fragrant and very perfectly shaped rose, of an exceedingly delicate blush color.

*Souvenir de la Malmaison*—perhaps the most perfect of roses. How I envy the grower who first saw that plant bloom, the seed of which he had sown, feeling that such a gem was *his*! It is magnificent either in bud or bloom. When well grown and fully expanded it is of immense size and most perfect form; color, pale flesh, slightly tinted with fawn, and brilliantly glossy. It is a robust, free grower.

Bourbon roses require but little pruning. Towards the end of February their shoots may be thinned, and the long ones shortened in proportion to their strength.

THOMAS AFFLECK.
[To be continued.]

———————◆———————

**The Rose—№ 6**

December 4:
The Rose—№ 6.
[Continued.]

The Southern Nurseries,
Washington, Adams County, Miss., October 1856.

CHINA ROSES—The ever-blooming habit and brilliant colors of nearly all of this class make it a favorite with all lovers of the roses. They are, however, in a great measure, without fragrance; but are pretty hardy, summer and winter, and require but little care to cause them to bloom most freely; though they richly repay a proper degree of care and attention. Amongst the best, or those which have thus far thriven best with me, are:

*Abbé Maillard*— A very showy and attractive, deep rich crimson colored variety.

*Agrippina* or *Cramoisi Superieur*—Is one of the [most?] showy and desirable of this, or, in fact, of any class of roses. It is a strong grower and most constant bloomer. Color, rich, brilliant crimson; large; beautifully cupped; each petal or flower-leaf has a delicate but clear white stripe down its centre.

*Clara Sylvain*— Pure white, large and excellent.

*Eugine Beauharnais* [*Prince Eugène de Beauharnais*]—A distinct, brilliant and beautiful rose, producing perfectly globular blooms of a bright amaranth [i.e., mauve and mauve-blend] color.

*Fabvier*—Scarlet; another brilliant and showy semi-double rose; very desirable for its bright color.

*Green Rose*[32]—Very curious; blooms freely; each bloom being a very double mass of dark green leaves. According to the newspapers this rose seems to command enormous prices in Europe.

*Indica Superba*[33]—Somewhat resembles the old *Indica,* or "daisy rose," so common here; but is a great improvement upon it; color, very pretty rose, paling towards the centre; very double and perfectly formed; perhaps the first to give good blooms in the spring.

*Mrs. Bosanquet*[34]—Pale flesh and very beautiful; habit quite vigorous.

*Nemesis*[35]—Very dark velvety crimson; rich and distinct.

*Prince Charles*—Beautifully capped, large, globular flowers of a brilliant, carmine color.

All of this class are largely employed in Europe to form beds or groups in the lawn; one color only being planted in a bed. The effect produced is exceedingly brilliant and beautiful. *Agrippina* makes a showy division hedge, and with care would make a very perfect fence.[36]

They require little more in the way of pruning than to cut out the weak spray, and shorten back the strong shoots to induce a growth of new wood, which alone bears the blooms. When planted in beds or groups, they are kept pegged down close to the ground, so as to cover the entire bed.

TEA SCENTED—The Tea roses are the greatest favorites of all with the ladies. Their extreme, but delicate, beauty, and rich, delicious fragrance, place them above all the others, and especially in the South, where they flourish

32. *Rosa chinensis viridiflora,* introduced by American John Smith in 1827.

33. There are several varieties of the China rose *Rosa indica.*

34. Also known as *Mistress Bosanquet,* introduced 1832.

35. Also given as *Némésis,* introduced in 1836.

36. As previously mentioned, Affleck proposed using hedges of roses (particularly Cherokee or Chickasaw roses) as a cheaper alternative to constructing and maintaining fences on plantation property. Once established, such hedges were effective at keeping workers on the property and intruders out. He offered this horticultural service to clients for a fee and explored the subject in *Hedging and Hedging Plants in the Southern States* (1869).

well. They bloom more perfectly than any other roses in the autumn. The severe cold of the winter of 1856 was almost too much for young plants of this class that were unprotected. I lost the greater part of my stock of young plants, They were in the most perfect and full bloom two days before Christmas; and being in that growing state, the severe freeze of the following night destroyed the young, and greatly injured the old plants.

*Abricote*[37]—Is of a bright rosy fawn color, delicate and beautiful.

*Adam*—Very delicate rose color, large and splendid.

*Bougere*[38]—A distinct and beautiful large and fragrant variety, of a glossy bronzed rose color; a very robust grower, and constant bloomer.

*Cassio*—Is one of those tea roses whose full-grown bud is so lovely; color, delicate rose.

*Cels*—One of the freest blooming of roses, and of a rich, glowing blush. Does not, however, always open well; requires good culture, and clear, warm weather to give perfect flowers; then, most beautiful.

*Devoniensis*—Than this lovely rose, there is nothing of its kind more perfect. The flowers are of vast size, finely cupped; color, fine creamy white, tinted with rose. Like many of the best roses, it does not give perfect flowers upon young plants.

*Govbault*—A fine, robust growing plant, and a free bloomer; color, bright rose, finely cupped, buds very perfect and fragrant.

*Hardy*—Delicate but vivid rose color; flowers of large size; a fine variety.

*Josephine Malton* [*Joséphine Malton*]—Creamy white, shaded with fawn; very large and perfect; a striking variety.

[Indecipherable] *Panache*—Delicate straw color, shaded with rose, very pretty.

*La Sylphide*—One of my especial favorites—so very fragrant, yields such a constant succession of its large flowers, of so pretty a rosy buff color, turning to creamy white, and so strong a grower, making quite a tree.

*Lyonnaise*[39]—Pale flesh color; large and blooms freely; the half opened buds are early beautiful.

*Princess Hélène*—Light rose, with a pretty and peculiar tinge of yellowish buff; flowers large and globular; a very desirable variety.

37. *Abricoté*, introduced in 1836.
38. *Bougère*, introduced in 1832.
39. Perhaps *Belle Lyonnaise* (1854) or *Beauté Lyonnaise* (1851).

*Princess Marie*—Dark flesh color; large and fragrant; one of the finest of this group—though, like Cels, does not always give perfect flowers.

*Sofrano*—Buds, before expansion, are of a bright apricot color; flowers, when open, fawn, or rather more of saffron. A robust, hardy and excellent rose. The buds are very lovely.

*Stombiot*—Cream-colored in the buds, but opens pure white; large and beautiful.

*Souvenir d'un Ami*—Color, delicate salmon, curiously shaded with rose—very singular and pretty. Flowers, large and beautifully imbricated; habit, good and vigorous, and a free bloomer.

*Triumphe de Luxembourg*—Buff rose; very large. A fine old variety.

*Victoria Modesta*—An exquisite variety; light rose, beautifully shaded, very double and perfectly formed. The flowers open up very finely.

*William Wallace*—This is a fine, free blooming rose, growing vigorously; bright blush—large and beautiful.

This class is more delicate, generally, than any of the others, and will not endure a very hard freeze, immediately following growing weather, entirely uninjured. They should, therefore, towards the approach of winter, have the earth heaped up a little around their stems, that they might not be entirely lost in case of an unusually severe winter.

In the pruning, they require similar treatment to the China rose.

THOMAS AFFLECK.

[To be continued.]

———◆———

### The Rose—№ 7

December 7:
The Rose—№ 7.
[Continued.]

The Southern Nurseries,
Washington, Adams County, Miss., Nov. 1856.

NOISETTES—The original of this class was a seedling produced near Charleston, S. C., from the old musk rose, fertilized with the common China, and is named after its original grower. It now includes some of the most magnificent roses we have. I have a very superior collection of them, being especially favorites. They bloom afresh after almost every shower, and bloom early and late in the season.

I have now before me, this 5th day of November, a bouquet of absolutely perfect blooms, all but one of this class, and on the day before last Christmas they were in equal perfections. There are those pretty little gems, Aimée Desprez and Donna Marle; Solfataire and Chromatella—the first almost as deep and rich in color as the last; a cluster of Gerbe des Roses ("sheaf of roses") and most admirably varied; it is classed with the Bourbons, by the way, though with much of the vigorous habits of the Noisettes, combining just a dozen of absolutely perfect blooms; Blanche de Lait (not now, however, "white as milk" but with a delicate blush tinge in the centre; as is its wont in the fall); Angelique Clement in another vast cluster; Mrs. Siddons, whose buds are exquisitely beautiful; and Ellinor Bouillard, in another prodigious mass of half-opened buds; but I forget that there is a limit even to the extent of your columns, Messrs. Editors!

*Aimée Desprez,* a most showy and beautiful miniature rose, though a vigorous grower; covered with myriads of pretty, rose-colored blossoms, very double, about the size of a half dollar, and in immense clusters. The full blooms are even more perfect in shape and each petal is yet more beautiful by being richly striped with a much darker rose color.

*Angelique Clement,* a fine, free blooming sort, bearing immense clusters of very double dark rose colored flowers.

*Augusta*—This new American seedling, though a most beautiful rose, scarcely equals the glowing descriptions by which it was heralded. It resembles *Solfataire,*[40] but the petals are larger and better formed, though not so numerous; color a bright lemon. It has a delightful tea fragrance—a strong runner—foliage rich and beautiful.

*Blanche de lait*—Pure white, in large clusters, perfectly formed; a great bloomer and vigorous grower. A very desirable variety, indispensable for bouquets.

*Chromatella* or *Cloth of Gold*—Is a truly magnificent rose. A strong grower, and, when the plant has sufficient age and is well cultivated, a free bloomer. Bud of a rich cream color, and very large; when fully expanded

---

40. *Solfataire* was exhibited (and was a prize-winner) at the Massachusetts Horticultural Society in 1849; a modern name is unknown. As Affleck notes, it was confused in its day with *Augusta,* as noted in the *Magazine of Horticulture* 169 (January 1849): 238; C. M. Hovey, ed.

the color is a brilliant and beautiful yellow. Like some others of our finest roses, it does not bloom well when young; but when well established for two or three years, in a good soil and location, nothing can exceed the magnificence of its superb flowers.

*Donna Maria*—A perfect little gem; especially when worked upon stout plants of *Manettie,* as it is not inclined to form strong stems of its own. The buds are perfectly formed and cupped; of a delicate blush, paling to flesh color when fully expanded; it is then a miniature of Souvenir de la Malmaison. It is a constant bloomer.

*Ellinor Bouillard*—Is a fine strong-growing distinct rose, bearing large clusters of pretty light pink flowers.

*Fellenberg*—One of the finest crimson cluster-blooming roses, of a singular and pretty tint of color; of strong habit and very showy; a fine pillar rose.

*Jeanne d'Arc*—is a vigorous growing, pure white rose, which makes a beautiful pillar.

*Lactans*—I found this grown under glass [i.e., in a glasshouse], in quantity near Louisville to cut in winter, when its pure white buds were in great demand. It is there called "Magnolia rose," grows finely here, of course, out-of-doors.

*Lamarque*[41]—A well-known, rampant-running climber, forming immense shoots; and almost always covered with its large, superb, white slightly straw-colored flowers. Fragrant, and most desirable.

*Mrs. Siddons*—A strong-growing, free-blooming rose, bearing numerous clusters of delicate fawn-colored flowers, tinted with rose; its buds are perfectly beautiful.

*Ophirie*—Color, very singular, bright salmon, almost saffron; very beautiful; blooming in clusters; habit of growth very strong.

*Solfataire*—Bright lemon; a fine old rose; strong runner, very beautiful.

*Triumphe de la Duchère*—A very desirable rose, of a pale rose color, very large, and blossoms in vast clusters.

---

41. Introduced 1830. In his journal, architect Thomas K. Wharton records seeing this rose in New Orleans, on January 8, 1854: "Cold as it is I noticed in going to market a fine Lamarck [Lamarque] Rose climbing over a two story gallery front, its leaves quite green and fresh and covered everywhere with large, full blown, white Roses, and innumerable beautiful buds ready to take their place." *Queen of the South: New Orleans, 1853–1862: The Journal of Thomas K. Wharton,* ed. Samuel Wilson, Jr., Patricia Brady, and Lynn d. Adams (New Orleans: Historic New Orleans Collection, 1999), 8.

*Victorieuse*—Very pale blush; buds perfect in form and color; a pretty dwarfish variety.

The *Noisettes* require singular treatment in that advised for the two preceding classes. To have them bloom in perfection, bear in mind that they must be well manured at least once a year.

THOMAS AFFLECK.

[To be continued.]

————————◆————————

**The Rose—№ 8**

December 13:
The Rose—№ 8
[Concluded.]

The Southern Nurseries
Washington, Miss., Nov. 20, 1856.

GENERAL CULTIVATION AND TREATMENT.

Although the rose may be grown in the highest perfection in even the smallest dooryard or plot of garden, on the borders of the vegetable garden, or scattered over the lawn, provided the soil is deep and rich, and the plants receive liberal winter-dressings of manure; yet wherever it can be done, a separate plot of ground should be appropriated to a *Rosarium,* which should be located near the home, and easily and directly accessible from it. Let the entire plot be trenched, if possible, to the depth of three feet, and thoroughly manured and drained. (Full directions how to trench ground, have been given in the *Southern Rural Almanac.*)

The walks may then be laid off, and edged with dwarf-box. The beds should not be made too small: but may be say from sixteen to eighteen feet in width; that a row of the tallest and strongest growers may be planted in the centre; with another of those of more moderate growth on each side, giving a space of six feet between, and another space between the outside rows and the edging, of two and a half to three feet.

The entrance to the *Rosarium* may, very appropriately, be underneath an arch formed of one or more of the BANKSIAS; or pillars formed of some of those HYBRID PERPETUALS or NOISETTES, adapted to that purpose. The entire Rosarium may be enclosed with strong hedges of the shown CHINA roses, such as *Agrippina, Abbé Mullana,* &c.

In the centre a group of beds might be laid off, to be covered with some of the most brilliant of those best adapted to *bedding out,* as it is termed. For this purpose plants *on their own roots* are preferred, or those which have been *root-grafted.* The plants are set out, say three feet apart; and as they grow out, their branches are pegged to the ground, and so pruned and again and again pegged down, as to cover the ground completely. Nothing can be imagined more showy and beautiful than such beds judiciously managed. *Giant of Battles, Du Roi, Marquise Bocella, Agrippina,* and a host of others may be thus employed. In a handsome and well-kept lawn those beds are exceedingly ornamental.

*Pillars of roses* are formed by training some of those strong-growing *perpetuals,* &c., best adapted to the purpose, to stout stakes of some durable wood, and firmly planted; pruning and training in such a manner as to present a perfect pillar of roses when in bloom. When thus clothed, from bottom to top with bloom, they are most lovely objects in the lawn or open grounds, or lining the main walks of the Rosarium. *Baronne Prevost, Pius IX, Fellenberg, Lamarque,* &c. are adapted to this purpose.

I would like to impress upon the grower of roses that they need not expect a vigorous growth, and, as a consequence, perfect blooms, from plants in poor, uncultivated soil, covered with grass and weeds, and especially Bermuda grass. The soil *must* be deep, rich and well tilled.

The manures of the household, preserved in tanks, or in barrels sunk in the ground, and applied to roses in December or January, or in a much diluted state say with soap suds, which are excellent for the purpose, during dark and showery weather at any other season, will be found most excellent. Any good, strong, not too chaffy manure, however, will answer, laid on the surface around the plants, early in the winter. The rains carry down the strength of the manure to the roots. At the first opening of spring, have the surface forked slightly over, and again lay on two or three inches of rough manure to serve as a mulching, and leave it undisturbed until the August rains set in, when the soil may again be lightly formed over, and a mulching of tree moss laid around, and pegged down.

"What!" says one of my lady readers, "all of this trouble, this pruning, and pegging down, this training and manuring, to grow roses! Mine grow very well without anything of the kind." Perhaps so; they may exist, and even grow and produce blooms. But, dear lady, you have no idea of the gorgeous beauty of this Queen of Flowers, taking blooms produced under such treatment as your criterion.

But no matter how well you may treat your plants, of this I must apprise you, that "every rose has its season." By this I mean that the very choicest varieties, during some season, will bloom imperfectly; whilst others, from which less might have been expected, have been wondrously beautiful and perfect. And hence the necessity for a very considerable variety, in order to have perfect blooms at all times; and which variety should often include those of colors very similar.

And how difficult it is to describe the color of roses in words! What words will express the intense brilliancy of color, for instance, of *Giant of Battles, La Bedogere,* &c., &c. "Brilliant, glowing crimson, approaching to scarlet," are those I have used in my catalogue published in the *Southern Rural Almanac* for the coming year, to convey an idea of the color of the *Giant.* Yet it gives but a faint conception of the intense brilliancy of that splendid rose when in perfection. And so with almost every other. The same words may be really appropriate to the color of two or three different roses, and yet those colors be very distinct, the one from the other.

There is often an objection made to "budded roses," that is, those which are budded or grafted upon other sorts. This has arisen mainly from the imposition practiced by those traders who bring roses and other plants to our Southern towns and cities, every winter, from France and Belgium—the refuse of the large nurseries there; but which are here sold as choice varieties of any name, color or shade *that may be asked for!* Although thousands have thus been most grossly imposed upon, and the impositions repeatedly exposed, those men continue to bring in and sell vast quantities of such plants every winter.

Many roses grow and bloom infinitely better when worked (budded) upon the *Manettii* than when on their own roots. The *Rosa Manettii* is a most robust and hardy variety, which I find to be the best for this purpose, after trying many others. It grows luxuriantly upon soils where none others would thrive.

But I find that I have already extended this article, and must bring it to a close; trusting that I have fully replied to the enquiries of my fair correspondents.

THOMAS AFFLECK

# 8

## The Health and Management of Enslaved Plantation Workers

This article is from a medical journal published in New Orleans that was intended for distribution in ten southern states and California. Like other accounts from the past, it must be read with an understanding of the circumstance in which it appeared. The *Southern Medical Reports* aimed to publish articles related to medical issues of the day, among them discussions of the racial differences between Negroes and whites, how those differences compelled Negroes to "servitude," and, it followed, to certain diseases and medical conditions. Some medical professionals claimed, in papers published in the journal, that there were such differences. Others were apparently not convinced, including Dr. Erasmus Darwin Fenner (1807–66) of New Orleans, the journal's editor, even though he published papers supporting that position.[1]

The following article, dated May 19, 1851, contains first a parenthetical statement from Fenner, followed by Affleck's responses to ten "interrogatories." Similar to his earlier report to the Commissioner of Patents of 1848, Affleck responded in a straightforward, factual, and detailed way; unlike the Patent document, however, here Affleck includes the questions prior to giving his answers.

Like Affleck's description of the plantation garden as one that could add to the "health and comfort" of its workers, here is an account of the general living conditions of enslaved workers on a Mississippi plantation in the early 1850s as described by someone who recognized the benefits of a certain measure of health care for those who worked there. Moral, social, or legal questions of the South's slave-based economy were not part of that discussion.

—Ed.

---

1. Fenner was prominent in the New Orleans medical community and wrote *A History of the Epidemic of Yellow Fever, at New Orleans, Louisiana, in 1853*. His descendants became (and remain) prominent in political, financial, legal, and social circles in New Orleans. His son, Charles

## On the Hygiene of Cotton Plantations and
## the Management of Negro Slaves[2]

(The following interesting communication, from an unprofessional gentleman of fine talents and extensive experience, was kindly furnished us in reply to some interrogatories we addressed to him last winter.[3] Mr. Affleck is one of the most scientific agriculturists to be found in the Southern States, and is well known in this region as the author of the *Southern Rural Almanac*, a work abounding in useful information to *planting, gardening, botany, soils, climate, etc.* At present, we believe, his attention is devoted particularly to *Horticulture*, but he has been an extensive planter for a number of years. His efforts to introduce more system and order in the management of plantation affairs, by means of the 'Plantation Record,' a blank book which he has admirably arranged for the purpose, are worthy of all praise. Nothing can be better calculated than the faithful keeping of such a record, to add a charm to that homely primeval employment (tilling the ground) which is by far the most important pursuit in this *great and growing* republic. This paper, in connection with that of Dr. Cartwright,[4] is but the beginning of a series, to appear from year to year, which, we trust, will promote the true interest of the master, and ameliorate the condition of the slave. We commend Mr. Affleck's paper to the careful perusal of our readers, both North and South.—ED.)[5]

---

(1834-1911), a Confederate officer, was later a justice of the Louisiana Supreme Court and active in erecting Confederate monuments around the city. Jefferson Davis died in the Fenner home in 1889. His grandson, also Charles (1876-1963), formed a brokerage house in New Orleans that later evolved into the national brokerage house of Merrill, Lynch, Pearce, Fenner and Beane.

2. From E. D. Fenner, M.D., ed., *Southern Medical Reports 2* (New Orleans: D. Davies, Son & Co., 1851), 429-36.

3. Affleck was "unprofessional" only in the sense that he was not a medical professional.

4. Samuel A. Cartwright (1793-1863), a physician who practiced in Louisiana and Mississippi, advanced pseudoscientific theories and published medical papers that reinforced the views of those who defended slavery. Among the "illnesses" he proposed in the early 1850s were "drapetomania" (the desire to flee from servitude) and "dysaesthesia aethiopica" (the apparent lack of work ethic among slaves). In his article "The Diseases and Physical Peculiarities of the Negro Race," which preceded Affleck's in this medical journal (pp. 421-29) and is a summary of a paper given earlier in New Orleans to the Medical Association of Louisiana, Cartwright enumerates anatomical and physiologic differences between whites and Negroes, claims that medical science has ignored these differences, and calls for a "radical reformation" in medical education.

5. This note is from Fenner, the journal's editor.

*Dr. E. D. Fenner, Editor Medical Reports:*

*My dear Sir:*—In reply to your inquiries relative to the health and management of negroes, and the sanitary condition of cotton plantations, I would remark, that my observations have been confined, more particularly, to the hill country embraced in this and the counties immediately adjoining; that they extend through a series of over nine years; and that my replies will be limited to the condition of things in the district referred to.

The face of the country is much broken; not, as is generally the case in hill regions, divided into valley and hill; but is sharply rolling, the tops of the hills representing the general surface, as it were—the hollows being depressions, forming a very peculiar character of country. Some few creeks and streams of water occur, but even their valleys are of limited extent.

The soil was, originally, very rich; but where the timber has been destroyed and the land tilled, the soil has been, in a great measure, washed off[6]; accumulating in the hollows; raising the level of the creek bottoms; and adding to the formation at the Balize.[7]

The general growth of timber in this country was originally oak in some eight or ten of the noblest species, elm, hickory-poplar (*Liriodendron tulipifera*),[8] mulberry, magnolia (*M. grandiflora, accuminata, cordata,* and *macrophylla*)[9] ash, linden, black walnut, sweet and black gums, sassafras, etc. In the low grounds, cypress, sycamore, etc. In some parts of the country there are scattering pines, occasionally running into pine-woods. In others the beach [i.e., beech], and again the magnolia preponderates, intermingled with the holly and wild peach.

In some parts of the county[10] springs of the most limpid water abound; forming *spring branches,* as they are termed; and emptying their waters,

6. As elsewhere in the South, soil erosion was a significant problem in the southwestern counties of Mississippi when forest coverings were removed. Kudzu, introduced to America in the late nineteenth century, was later used in Mississippi and other southern states in misguided efforts to control this soil erosion.

7. Affleck's meaning here is unclear, but we may infer that his reference is to sedimentation at the terminus or delta of a creek or river. Fort La Balize was one of the first settlements founded in America by the French, at the mouth of the Mississippi River. Its importance was due to its geographical position, enabling those who lived there to control passage up-river, and its name means sea mark, a maritime beacon or marker for navigation purposes.

8. I.e., tulip tree, tulip poplar, or poplar.

9. Respectively, southern magnolia, cucumber tree, yellow cucumber tree, and big-leaf magnolia.

10. Presumably Adams County, Mississippi.

principally into St. Catherine's and Second creeks. These springs break out of beds of silicious sand or gravel, and occasionally from the beds of *marl*, which abound in many parts of the county.

The country having been long settled, the land is very much exhausted; not a few plantations [are] on the verge of abandonment for the rich lands of the Louisiana swamps, where more cotton may be made, though with less clear profit in the end, and with a vast amount of suffering and toil to white and black.[11] The number of negroes, in the country has greatly diminished, I suspect, during the last ten years, from the removal of numbers to other districts.

With these preliminary remarks, I will proceed to reply to your inquiries in detail.

1st. 'What is the customary method of feeding, clothing, housing and working negroes?'

Their *food* is cooked for them by one appointed for the purpose and directly responsible to the master or overseer. They have as much well-cooked food as they can consume; the general allowance of meat being 3½ to 4 pounds per week, of sound mess pork; or its equivalent in bacon, to each working hand over, say, 10 years; with bread, hominy, vegetables, etc., *ad libitum*. Some few, to their shame be it said, do not feed so well; but they form the exception, and are themselves greatly the losers thereby. Fish and molasses are given occasionally. Not nearly enough of vegetables are grown and fed to negroes. Each negro has his small tin bucket, with a cover, in which each meal is sent out by an old man and cart; breakfast at 7 to 8 o'clock, dinner at 1 to 2, and supper when they drop work. The buckets are washed after each meal. Water is either hauled out in a barrel on a sled, or got from springs, etc., and is carried about amongst the hands at work as wanted, by youngsters.

*Clothing* consists of, for winter, a roundabout or loose coat, a pair of trousers of strong woolen jeans, and two shirts of stout lowells, for the men; and for the women, a frock of warm linsey, and two chemises of lowells[12];

11. Throughout his career, Affleck was a proponent of rebuilding exhausted soils through various means, such as fertilizer, crop rotation, and allowing fields to turn fallow. Many farmers were resistant to trying these "scientific" techniques, preferring instead to abandon exhausted soils and move westward, a practice that accelerated population growth on the frontier but left exhausted and depleted soils behind.

12. Lowell cloth was a generic term for inexpensive, coarse cotton cloth, produced originally in the textile mills of Waltham, later Lowell, Massachusetts, and eventually on plantations and

our winter being short, their last year's suit is seldom nearly worn out; and youngsters who are very hard on their clothes get additional trousers, etc., when necessary. One pair of strong russet brogans, and a hat or cap of some kind. One year, a warm capote or twilled blanket; and the next a bed-blanket for every negro, even to the infants. Some make comforts* for their people occasionally. The children, too, have comfortable clothing.

Housing is almost always good. Each family has a room from sixteen to twenty feet square, many with an additional shed-room, and all with galleries and porches. The houses are of brick, frame or log, some of pisé.[13] Frame or log, generally, considered most conducive to health. Each house has its yard, with poultry houses, etc., and some with gardens.

The *working* is quite uniform. At day-break the bell rings or horn blows, and all hands turn out; the mothers carrying their children to the nursery, and all proceeding to their work, so as to commence by sun-up. During the winter, from half to three quarters of an hour is allowed for each meal; during summer, half an hour for breakfast, and from one and a half to two hours for dinner. All hands quit at dusk. Some planters do not drop soon enough, nor arrange their work with that system, to admit of their hands being ready for bed at a sufficiently early hour; but they form the exception.

2d. 'What are the principal diseases from which negroes suffer in your region? and which are the sickliest seasons of the year?'

During *winter*, principally *pneumonia*, occurring in wet and cold weather, and greatly aggravated by the unskillful treatment of overseers; warded off by dry, well-aired houses, warm and sufficient clothing and food, and an avoidance of exposure to wet, which the negro cannot be exposed to with impunity. In *spring*, few diseases; none, if dry; if wet, apt to be sickly— pneumonia, and intermittent fever; rarely subject to those bilious attacks to which the whites are liable at this season. *Summer* usually very healthy; some bilious fevers in June; but when fodder-pulling begins, negroes become sickly, being decidedly the most injurious work to their health that is done upon plantation, from their exposure to heavy dews, over-heating

---

textile mills in the South. Linsey-woolsey is a coarse, plain, sturdy woven fabric of wool and linen or cotton. For accounts from former slaves of these materials, see http://library.uml.edu/clh/All/Lowcl.htm, accessed July 26, 2012.

13. An ancient and universal technique of building using the compressed raw materials of earth, chalk, lime, and gravel.

amidst the tall corn, and when reaching the end of the row, drinking large quantities of water. In the *fall,* violent congestive fevers are occasionally produced during cotton-picking, when cold nights set in, accompanied by hot days. This is greatly lessened by the practice, now generally adopted, of requiring them to change their wet clothes for dry, so soon as the dew passes off. When the cotton-leaves fall, sickness lessens. Upon the whole, I have never seen an equally healthy region, for white or black; and my observation has been somewhat extensive. *Cold and wet winters* the most dangerous sickly seasons; diseases of no kind prevailing during dry seasons, or warm and dry summers.

The *principal causes of sickness* upon the plantation, are the use of spring, well, creek or bayou water,—(it is a *fixed fact,* that cistern or rain water alone is healthy; instances being quite common, where places notoriously sickly, though supplied with abundance of pure, clear and cool spring water, becoming at once equally healthy from the exclusive use of that from underground cisterns: there should be large cisterns, not only at the house and quarter, but at the gin-house[14] and weather-shed in the fields)—[for] night work and night rambles, coon hunting, etc. There is nothing gained by fagging[15] the hands with corn-shelling, carrying seed cotton into the gin, etc., after night. They should drop all work in time to have a couple of hours before the bell rings for bed, and for seven to eight hours of uninterrupted sleep; and should have Saturday afternoon, whenever work is not unusually pressing, for washing, house-cleaning, tending their crops, etc., the overseer making it his business to see that they are so employed. Badly-cooked food, wherever permitted, an insufficiency of vegetables, and a want of cleanliness, are all causes of sickness. Much injury is frequently occasioned by the hands carrying their basket full of cotton, during picking, for any great distance, on their heads. A load of 100 to 150 pounds pressing upon the skull, neck and back-bone, when the muscles are relaxed by fatigue, cannot but be injurious, and *is* a decided cause of sickness and accidents, such as sprains, ruptures, etc. In every instance, additional care in food, clothing, and household comforts; a ready supply of fuel in cold weather; an avoidance of exposure to rain and night air and dews; strict discipline; reasonable hours and

14. A "gin-house" was where cotton was "ginned"—the act of separating the fiber from the seeds by a mechanical engine or "gin," giving rise to the term and its usage.

15. "Fagging" is to devote serious or sustained effort to an endeavor.

moderate punishments, are followed by a corresponding degree of health and strength, and increase in the numbers of negroes.

3. 'Are the whites and blacks equally liable to the customary prevailing diseases?'

Negroes have diseases peculiar to themselves; and even in the same diseases, the symptoms, etc., are different, and the treatment must also be very different. I think them equally liable, under the same circumstances of food, exposure, etc.

4. 'What influence does acclimation appear to have upon negroes brought from the most northern States?'

The first year, no marked influence; the second summer, extremely liable to dangerous attacks; after that, the acclimation seems to be complete.

5. 'What is the comparative duration of life among whites and blacks, creoles,[16] and immigrants?'

Cannot well say; the negro outlives the white or the mulatto. I am inclined to think that, after acclimation, the more active out-of-door habits of the white immigrant, is conducive to health and duration of life.

6. 'What seem to be the principal causes of disease among negroes, in respect to food, clothing, exposure, filth, water, drink, etc.?'

Already answered, in a great measure.

7. 'Are negro women, under the ordinary regime of plantations, as prolific as white?'

Yes: more so, when not over-worked. As a general thing, decidedly so.

8. 'What is the ordinary management of negro children? And what the comparative mortality between whites and blacks?'

Upon every plantation working, say, twenty hands or more, there is a nursery, with a careful old woman, whose business it is to take care of them, wash them, cook for them, etc. Every mother carries her cradle, blankets,

---

16. Here, creole refers to those of mixed race; another term is mulatto, used later. In early-nineteenth-century New Orleans and before, Creole referred to those born in America or in the New World to parents of European, mainly French or Spanish, heritage. Later, the term evolved into other meanings: anyone of mixed race, regardless of heritage, as used here; or as a reference to free people of color. Over time, the term and whether or not it appears with an upper case C has remained ambiguous, depending on the context of its use and user. See Arnold R. Hirsch and Joseph Logsdon, eds., Creole New Orleans: Race and Americanization (Baton Rouge: Louisiana State University Press, 1992), and Sybil Kein, ed., Creole: The History and Legacy of Louisiana's Free People of Color (Baton Rouge: Louisiana State University Press, 2000).

mosquito-bar and child to the nursery, before she goes out in the morning, after suckling. Whilst under nine months, the aim is to have the child suckled every three and a half to four hours, the mother coming in for the purpose; and this rule is rarely exceeded. Where the distance is great, say over three or four hundred yards, good managers have a large, dry, airy shed in the field where the hands are at work, and there the children, cradles and all, are taken by the nurse, in a cart, or otherwise, so soon as the dew is off— an excellent practice. The mortality of negro children is as two to one when compared with the whites, depending solely upon locality and care. Quarters are often badly located; children allowed to be filthy; are suckled hurriedly, whilst the mother is over-heated; and laid on their backs when mere infants, on a hard mattress, or a blanket only, and rocked and bumped in badly-made cradles; not a few are over-laid by the wearied mother, who sleeps so dead a sleep as not to be aware of the injury to her infant; a vast proportion die under nine or ten days, from the most unskillful management of negro midwives, who do not know how to take care of the navel, and doses the infant with nasty nostrums from the moment of its birth; from having access to green fruit, eating acorns, etc., and from dirt eating. Of those born, one half die under one year; of the other half, say one-tenth die under five years; and of the remainder, a large proportion are raised. Dirt-eating is frequent amongst young negroes, and always kills them, if not cured.[17] The constant use of molasses is said to induce it, but I cannot say how correctly. Those under the best care are liable to it. Seems to be occasioned by a morbid state of the stomach, and should be so treated. One dirt-eater upon a plantation will infect the whole. Mostly infected at from two to ten years, say one child in forty eats dirt. Children should have no sweet milk; none but sour, or buttermilk. They are very liable to worms, which kill a good many, or stunt them: cured by giving every child a spoonful or two of a strong, sweetened

17. Dirt-eating, geophagy, was once thought to be the body's response to hunger and adding (or replacing) needed minerals. A recent study at Cornell University, led by an authority on maternal and child health and nutrition, found that the practice has little to do with being hungry or seeking minerals the body might be lacking. Instead, it may help stave off existing intestinal pathogens, especially in pregnant women and pre-adolescent children. Pregnant women and children living in tropical regions of the world—the individuals who often consume dirt—are also the most vulnerable to parasites and pathogens. The practice of geophagy makes sense to these groups, especially if they're experiencing acute digestive illness. See http://www.human .cornell.edu/bio.cfm?netid=sly3and and http://news.discovery.com/human/eating-dirt-more -adaptive-than-thought-110607.html, accessed July 26, 2012.

decoction or tea of the root of the China tree (*Melia azederach*),[18] every other morning, till five or six doses are taken.

9. 'What is the customary medical attention devoted to sick negroes; and what care extended to the disabled and worthless?'

As a general thing, there is not sufficient good medical attention on plantations. Too much is left to the overseer, who doses after a routine—an emetic followed by calomel and oil. There are exceptions, of course. The disabled and worthless slaves are better cared for than the same class in any other community on earth.

10. 'What seems to be the state of feeling between master and slaves in good families and well-governed plantations?'

Invariably good, upon *well-governed* plantations. I am not aware of a difficulty of any kind having occurred in this county, since I have resided in it, between the negro and his master or master's family. Occasionally difficulties *do* occur between the negroes and their overseer; almost invariably the fault of the master, in trusting too much to the overseer, and too lax discipline.

In bringing these hurried replies to a close, I must remark that it could require an octavo volume of large size to contain anything like full answers to your questions. I have done the best I could with the leisure at my command.

It is extremely difficult to arrive at correct sanitary and statistical results, anywhere in the South, from the want of correctly-kept plantation records. Planters are not by any means fully advised of the importance of such records; nor are they sufficiently stringent in requiring them to be kept by their overseers, on those plantations on which they do not themselves reside; nor in keeping them regularly and correctly where they do not require it of the overseer. A uniform system of plantation records would direct the attention of the community to the most prevalent evils, and to the best practices, alike; so that the one would be avoided, and the other generally adopted.[19]

Yours, very truly,<br>THOMAS AFFLECK.

Washington, Adams Co., Miss., 19th May, 1851.

*A quilted coverlid [i.e., coverlet—Ed.].

18. Also known as China ball or chinaberry.

19. The reference here, of course, is to Affleck's own *Record and Account Book* for both cotton and sugar plantations. Note, however, that he mentions the need but does not suggest his publications as a solution.

# 9

## On Ornamental Trees and Shrubs

Affleck, unlike his contemporary A. J. Downing, is not known today as one who wrote about design or gave design advice to his readers. However, like Downing, he did have a sense of how plants might fit in the landscape and how they could be used for decorative or functional effects. From that perspective, one of Affleck's most interesting articles is the following from the Natchez *Courier* on the value of ornamental trees and shrubs for urban situations. Here we see how his knowledge of trees, including those native to the region, informs suggestions he had for making spaces more attractive.

His suggestions are aimed mainly for domestic uses ("there is a sameness in our lawns and dooryards"), but he does mention public spaces ("your streets and pleasant Bluff promenade") as locations for tree plantings. Since this article appeared in the Natchez newspaper, Affleck was referring here to the public park in Natchez on the bluff overlooking the Mississippi River. By coincidence, Olmsted had visited this site just months before Affleck's article appeared, and made the following observations:

> But the grand feature of Natchez is the bluff, terminating in an abrupt precipice over the river, with the public garden upon it. Of this I never had heard, and . . . I strolled off to see the town, I came upon it by surprise. I entered a gate and walked up a slope, supposing that I was approaching the ridge or summit of a hill, and expecting to see beyond it a corresponding slope and the town again, continuing in terraced streets to the river. I found myself, almost at the moment I discovered that it was not so, on the very edge of a stupendous cliff, and before me an indescribably vast expanse of forest, extending on every hand to a hazy horizon, in which, directly in front of me, swung the round, red, setting sun.

He continues, describing the impact of the river ("though the fret of a swelling torrent is not wanting, it is perceptible only as the most delicate chasing upon the broad, gleaming expanse of polished steel") and the character of the

"park" ("some shrubs") and its inhabitants ("a man, bearded and smoking, and a woman with him . . . were the only visitors except myself and the swine").[1]

Certainly Olmsted's writing is more evocative of the landscape in general than Affleck's, and Olmsted's observations predict his later abilities to understand and manage projects on a grand scale, from Central Park at his career's beginning in the mid-1850s to the Biltmore estate grounds at the end, in the early 1890s. Affleck's writing is more focused on individual elements and the specific characteristics of each plant he discusses, but there is little evidence of any designed landscapes or surviving plans from Affleck's hand to know how he would have applied such knowledge in a design situation. The differences between each man's approach define the career of each and explain why Olmsted is well known and Affleck remains obscure. While both Olmsted and Affleck had an appreciation for plants and what they can do, their interests lay at opposite ends of the spectrum in terms of focus. That dichotomy has defined the profession of landscape architecture from Olmsted's career to the present and continues to be a conundrum that the profession has yet to reconcile.

1. Olmsted, *A Journey in the Back Country in the Winter of 1853-54*, 37-38.

## Article from the Natchez *Daily Courier*[2]

HORTICULTURAL

Washington, Adams County, Miss.,
October 24, 1854

*Editor Natchez Courier:*

Dear Sir—In common with your many readers, I have had much pleasure in perusing the very interesting and valuable articles on "Fruit-growing in the South," by RUSTICUS.[3] The information they contain was just of the kind we most needed.

As your "City of the Bluffs" seems to have become greatly alive to improvement, of late years, and many neat and home-like houses have been erected in and around the city, a few hints on planting ornamental trees and shrubs, with short descriptions of some of the less common and rarer sorts, may be apropos and useful.

We lack *variety*, as a general thing, in this class of dress and plants. In a climate in which a greater number of rare and extremely beautiful evergreens are perfectly hardy, than in any other I know of, unless perhaps the Isle of Wight, off the South coast of England—and doubtful if even there—we confine ourselves to some half-dozen kinds. Nothing can be more beautiful than the Laurier Amandier (*Cerasus caroliniensis*), Cape Jessamine,[4] Arbor Vitae, some of the Viburnams, Pittosporums, Euonymus, and Myrtles; yet, there is a sameness in our lawns and dooryards, from the general and almost exclusive use of these, that might readily be relieved by the addition of some of the many others which are equally, and in some instances, more beautiful.

---

2. This article subsequently appeared (in two installments) in *De Bow's Review and Industrial Resources, Statistics, etc.* . . . (New Orleans and Washington City, 1855) 19: 717–19, and 20: 214–18. Its earlier appearance in the Natchez *Daily Courier* was not noted in *De Bow's*.

3. Here Affleck is referring to articles contributed to an unnamed publication by a correspondent writing as "Rusticus." The publication could have been *The Soil of the South: A Monthly Journal Devoted to Agriculture, Horticulture and Rural and Domestic Economy*, from Columbus, Georgia, starting in 1853, which refers to "our correspondent 'Rusticus,'" who had contributed a tribute upon the death of Andrew Jackson Downing (Vol. III, 1853, p. 409). Affleck wrote a letter to the editor (Dec. 24, 1856) of this journal concerning the founding of a Southern Pomological Society that had appeared in Vol. I (1853, pp. 58–59). In 1857, *The Soil of the South* merged with *The Cotton Planter* in 1857 to form *The American Cotton Planter and Soil of the South*, published in Montgomery, Alabama.

4. Laurier amandier (*Cerasus caroliniensis*) is a laurel, akin to cherry laurel (*Prunus caroliniana*). Cape Jessamine is likely cape jasmine (*Gardenia jasminoides*).

So with our shade trees. The perpetually recurring Pride of China tree [i.e., chinaberry, *Melia azedarach*], beautiful though it be, to the exclusion of the scores of magnificent trees, native and introduced, is, to say the least of it, in very bad taste. It is a filthy tree, too, about a yard, when compared with many others.

As a shade and ornamental tree, there is none will compare with our magnificent Water oak [*Quercus nigra*], and Live oak [*Quercus virginiana*]. The latter is the more beautiful and permanent; the former is of somewhat more rapid growth. Suppose that, instead of the China tree, your streets and pleasant Bluff promenade had been lined and shaded with these oaks! By this time, you would have had ornamental trees such as few cities can boast of. The Mobilians [residents of Mobile, Alabama] were alive to the beauty of the Live oak as a shade tree for their streets and squares, and see the result now!

The Cork-oak (*Quercus suber*), the Holly-leaved [holm oak, *Quercus ilex*] and the Cut-leaved Turkey-oak [*Quercus cerris*] are all very beautiful, though yet somewhat rare. I have fine young trees of all of them.

The Imperial Paulownia [*Paulownia tomentosa*][5] with its immense leaves and numberless spikes of blue bell-like blossoms, has been introduced some ten or a dozen years, and is quite an acquisition. It blooms here, abundantly, both spring and fall.

The Varnish tree (*Stericlea platynifolia*)[6] is so called from its beautiful glossy bark, and large rich colored leaves, which seem all to have been recently coated with green varnish. It is, altogether, a pretty and desirable ornamental shade tree.

The Croton tree, and Everblooming china are both pretty trees, though, in a severe winter, the ends of the branches are sometimes killed by the frost.[7] The *Acacia julibrissin,* or flowering Acacia [*Albizia julibrissen],*[8] though by no means rare, is yet too showy, with its myriads of pink and yellow flowers, to be omitted in pleasure grounds, or even small yards.

5. Also known as princess tree, an exotic from China.

6. Now *Firmiana simplex,* Chinese parasol tree or Japanese varnish tree. Introduced in the mid-nineteenth century, this tree was considered exotic and desirable.

7. Croton is the common name of a large group of plants in the Euphorbia family, commonly evergreen shrubs; in Africa there are varieties that become small trees. It is unclear what the "Everblooming China" tree is.

8. *Albizia julibrissin* is the mimosa commonly found in the American South; it has clusters of pale pink, powder-puff-like flowers. *Acacia dealbata,* also known as mimosa or silver wattle, has yellow blooms.

Several of the Maples are natives here, and form, as elsewhere, most beautiful trees. Perhaps the best of these is the Scarlet Maple [Swamp red maple, *Acer rubrum drummondii*], so showy in the spring with its bunches of bright scarlet blossoms. The ash-leaved Maple (*Negundo*) or Box Elder, cannot be excelled as a shade tree in any country, where it has room to grow and spread. Several of the European Maples do well here, and are desirable trees.

The Chestnut is one of the most stately trees of the forest, and desirable not only as a lawn tree, but for its fruit. The large fruited Spanish is the finest.[9]

Our Great Southern Cypress (*Taxodium*) should never be omitted, where the soil is rich and moist. The chief cause of its rarity in lawns, etc., is the difficulty of transplanting young trees from the swamp to the dry upland of our hills. With trees grown on dry land from seed, there is no such difficulty.

The graceful weeping willow, though so easily grown, is comparatively rare. The curled-leaved variety, being quite as *weeping* in its habit as the other, is very curious. Each leaf is curled up like a cork-screw.[10]

The Ginko (*Salisburia*) [now *Ginkgo biloba*] or Maiden-hair tree is pretty, and quite ornamental. The leaves are very curious.

The Double-flowering Peach [*Prunus persica*] is one of the most showy of trees, forming early in the spring, a mass of wreaths of rich and extremely double, rose-like blossoms.

Where there is room for a few large and wide-spreading trees, the Pecan should not be overlooked [*Carya Illinoinensis*, a native]. They afford a fine shade, and come into bearing in eight or ten years. We know of one gentleman in Western Texas who has some 15 or 20 varieties of this delicious nut, which he has succeeded in multiplying by grafting. Two years ago, he sent the writer a quantity of nuts from each of 8 or 10 of the finest of his selections. These were planted, and have produced a fine lot of trees; the trees from each variety of nut show a wonderful family likeness, in foliage, habit of growth, &c.; whilst there is a marked difference between the lots. They have been all twice transplanted, and root-pruned each time; thus in a great measure obviating the difficulty in transplanting when the trees are older.

9. American chestnut (*Castanea dentata*); Spanish chestnut (*Castanea sativa*).

10. Weeping willow (*Salix babylonica*); corkscrew or curly willow (*Salix matsudana var. pekinsensis*).

The Mountain Ash, or *Rowan tree*, dear to every Scotchman's boyish recollections, we have succeeded in acclimating.[11] It is a beautiful tree.

The large-leaved Magnolia (*M. macrophylla*) with that same difficulty of transplanting from the woods, is quite rare in our gardens; where its magnificent foliage and immensely large and showy flowers fully entitle it to a first place. When grown from seed in the nursery-row, there is no difficulty in removing it.

Of Evergreen Shade Trees, the *Magnolia grandiflora* stands first. Like its companion the *Holly*, it is not easily removed from the woods. When quite young this may be effected, by lifting with a ball of earth around the roots, in the spring, and cutting off the leaves, but leaving the leaf-stalks. They well deserve that every available means should be used to secure both—the Magnolia and the Holly (*Ilex opaca*)—wherever shade and ornament are sought for. During the first three or four years from the seed, their growth is quite slow; but afterwards they push up rapidly, and soon form handsome trees.

There is another Holly, a native of the South and an evergreen, that is very generally overlooked. It is more commonly planted about Mobile than anywhere else. This is the *Ilex Vomitaria* [yaupon]. The growth is slender, leaves small and numerous, and in winter the plant is covered with bright scarlet berries.

Of the various *Coniferae*, it is rare to find a plant in a lawn in all this region; unless, perhaps, an occasional Long-leaved or Old-field Pine—both most noble and beautiful trees, and not planted one for a thousand that should be. There are many other Pines, from all parts of the world, now to be found in the nurseries, and all desirable.

The Spruces are the most prized of this family in Europe, although so common, that they are planted by the thousand to serve as screens to lawns and gardens, and to plantations of other less hardy trees. The Norway Spruce (*Abies excelsa*)[now *Picea abies*], the most common, is also the most beautiful. In fact, I know of no tree that equals it in gorgeous and impressive beauty. Some ten years ago, I imported a lot of fine plants of this and other spruces; and, as in every other attempt to import young evergreens either from the North or Europe, I saved but a very small percentage. Of those

---

11. *Sorbus aucuparia,* European rowan; *Sorbus Americana,* American mountain-ash; a small deciduous tree with showy white spring flowers and fall berries, used for medicinal purposes. Affleck obviously recalled this tree from his youth in Scotland.

saved were two Norway Spruces. For five years they did not make a growth of more than an inch each year! After that they shot up rapidly, and are now beautiful, healthy plants, eight or ten feet in height. Since then I have been more successful in habituating young plants to the climate, and have fine young trees of several species of Spruce.

The Cedars are very beautiful. And, by the way, what we know as the Red Cedar, is a Juniper, bearing a small purple berry; the Cedars are cone-bearing. Cedrus deodara, the Great Indian Cedar, is the most splendid tree of this family; perfectly hardy here, and of very rapid growth; rare, however. The Cedar of Lebanon is also hardy, but of much slower growth.

Two new evergreen Conifers, Cryptomeria Japonica and Cunninghamii Sinensis—the former from Japan, the latter from China—I look upon as great acquisitions.[12] Both are at home in our climate; requiring, however, like all of these resinous evergreens, a light and sandy, but rich soil; and are most graceful and beautiful, yet curious ornaments to the lawn or door-yard.

Another of these, the great Chili Pine (Araucaria imbricate) has not succeeded so well; though I have now a few young seedlings that seem to feel themselves at home.[13]

The Junipers, headed by our own beautiful native, the so-called Red Cedar (J. virginiana,) are indispensable. In the "Red Cedar" there is a great diversity of foliage and habit of growth; some being open and loose in habit, others upright and compact. The latter I have always selected from the seed-bed. They should have room to grow, and be allowed to sweep the ground with their branches; not pruned up into the likeness of a gigantic broom!

The Swedish Juniper [Juniperus communis] is a very upright in growth, and with fine and delicate, silvery foliage, and altogether a pretty plant.

The Arbor Vitæ (Thuja occidentalis) is well known—that is, the Chinese, (orientalis) the sort common here. And to form a pretty screen hedge, I know of nothing more beautiful; requiring to be kept nicely clipped, and the seed comes picked off so soon as large enough—otherwise the foliage becomes brown.

The American Arbor Vitæ (Thuja occidentalis) is still a more desirable plant; bearing the shears equally well, retaining its color better, and the

12. Cryptomeria Japonica, Japanese cedar; Cunninghamii Sinensis could be Cunninghamia (China-fir; Chinese cedar), Cunninghamia lanceolata.

13. Known in the nineteenth century as the Chili (now Chile) pine, also known as the monkey puzzle tree; native to southern Chile and western Argentina.

foliage giving out a sweet odor when crushed. The *Thuja plicata* is a wavy-foliaged, pendulous kind, also quite pretty.

It was long before I succeeded with the *Yews*. The English Yew is now perfectly healthy and grows vigorously. Its close, dark green foliage renders it very desirable, and especially in the cemetery, where from time immemorial, it has been considered the most fitting ornament.

And so with the Tree Boxes—the neatest and prettiest of evergreen trees; always fresh and pleasant to look on. They grow better here than even in their native climate; as does, also, the Dwarf Box [*Buxus sempervirens; Buxux sempervirens Suffruticosa*] for edgings.

The *Euonymus,* evergreen and variegated, are both very ornamental. The evergreen is often misnamed Tree Box. They are very hardy and grow rapidly.

There are several of the *Viburnums* which are handsome evergreens. *V. lucidum?* has rich dark foliage and showy white flowers, and makes a large plant. *V. laurus-tinus* [*Laurustinus Viburnum tinus "Lucidum"*], or Laurustinus, is one of the very richest of our flowering evergreens; blooming too, so very early in the spring, or in the winter rather, as to be very desirable.

The Laurels are all beautiful. But, like many of our finest plants—because not named in Northern books on gardening; and because Downing expresses his regret, at the same time that he gives expression to his admiration of the plants, that they are "too aristocratic in their nature to thrive in our Republican soil!"—the whole tribe has been overlooked. The *Lauris nobilis,* the Portugal, the English, and the Carolinian laurels, are perfectly hardy—the three first after *habituation* to the climate—and are rich and very beautiful evergreens. I have splendid plants of all, and especially of the English—(*Cerasus laurocerasus*).[14]

The *Photynias* [now *Photinia*] or Japan Hawthorns, are, like the Laurels, as yet somewhat rare in our gardens and lawns. There is a superb plant of the smooth-leaved Photynia (*P. glauca*) in Mr. Profilet's garden behind the Episcopal church in the city of Natchez, which has been for many years an object of admiration, and especially when covered with its myriads of snow-white blossoms. It is, I should judge, some twenty-five feet high, affording

14. Worth noting here are, first, Affleck's acquaintance with Downing's work and other "Northern books on gardening," and second, his disapproval of Downing's opinion. It is not clear where Downing's view about laurels first appeared. Leaves of the *Lauris nobilis,* laurel or bay laurel, are commonly used in cooking.

a fine shade. The Holly-leaved (*P. serrulata*) is yet more beautiful. I do not know of a richer evergreen. The small-leaved is also very pretty.[15]

The Pittosporums [*Pittosporum* sp.], both evergreen and variegated, are well known and favorite plants. They bear the shears well.[16]

*Olea fragrans,* the Fragrant Olive,[17] is an universal favorite with the ladies, and most deservedly so.

There are several of the Privets which form beautiful ornamental evergreen trees. The handsome evergreen, so generally admired, on the top of the mound, between the house of our friend Mr. Andrew Brown and the river, is the *Chinese privet* [*Ligustrum sinense*]. It is at all times a beautiful plant but more especially when covered with its racemes of white flowers. The Evergreen, the Myrtle-leaved, and Box-leaved, though commonly all used for hedging, may be readily trained into very pretty smallish trees.

The Japan Plum—*Eriobotrya (Mespilus) Japonica* [also known as loquat]—whether as a mere ornamental evergreen, for which it is second to few others; or for the fragrance of its flowers, or delicious fruit, is deserving of infinitely more attention than it has received. It has hitherto been somewhat scarce, and what few there were, were budded on the quince. I have now large, healthy and handsome trees, *seedlings grown here,* many of which are now full of blossoms. The Japan plum has ripened its fruit repeatedly in this county; and a very delicious fruit it is. It is now abundant in the markets of New Orleans in April.

The *Gardenias*—Cape Jessamine is the most common—are of course indispensable. The dwarf kind (*G. radicans*) [*Gardenia jasminoides*] is a lovely little plant. Fortune's new Chinese (*G. Fortunii*) was lauded so highly that I feared a disappointment. But it proved to be all he represented— the foliage larger and richer, and the blossoms fully double the size and more perfect in form; and though fragrant, not so oppressively so as the old sort.

But I have already extended my notes to such a length, that I must now be brief.

15. There are numerous photinias. Chinese photinia is *Photinia serrulata* (sometimes *serratifolia*). What is today known as Japan hawthorn is another plant altogether, *Crataegus cuneata*.

16. There are numerous varieties of pittosporum; Affleck is probably referring to *P. tobira* or *P. tobira* "Variegata."

17. Also known as *Osmanthus fragrans,* sweet olive.

I find I have omitted a very beautiful ornamental plant, which forms a small tree—a great favorite of the ladies—the Venetian Sumac, Fringe or Mist Tree; the blossoms appearing in numerous and delicately colored, haze-like spikes.[18] It is not evergreen, but a lovely plant in a group of evergreens.

The Deutzias, Spireas, Buddleas, Weigelas, Jessamines, Crape Myrtles, Double-flowering Pomegranites [sic], Forsythias, Honeysuckles, Lilacs, Snowballs, Shyringas (or Mock Orange), Ivy Brooms, with a host of other beautiful plants I must leave for another opportunity to describe.

As to transplanting shade trees, ornamental plants, and especially evergreens—bear in mind, that a thing that is worth doing at all, is worth doing well! And act up to it. Let the ground be properly prepared. If the entire lawn was well manured and thoroughly trench-plowed, and garden or door-yard well and deeply dug, so much the better. When this cannot be done, let large holes be dug, but not too deep if in a stiff retentive clay. For evergreens, provide a supply of rich, black, leaf-soil from the woods, and of well-rotted manure; for deciduous trees, any good and not too rank manure will do.

If your shade trees are to be procured from the fields or woods, select those only which grow *in the open*—not from dense woods or thickets. Spare no pains in the taking up all of the roots that can be saved, and especially the small fibrous ones; covering them with wet moss or gunny bags, or old carpets, etc., to keep them from being dried by the air or sun. When the tree is a handsome or valuable one, it should be lifted with a sufficient mass of earth to insure its safety. Trees or shrubs which have been prepared the year before for removal, as is done in all good nurseries, can be transplanted with infinitely less risk than those from the fields or woods. I have large specimens of English Laurel, Euonymus, Cedars, Junipers, Spruces, Japan Plums, Hollys, Magnolias, etc., which have been repeatedly root-pruned, so that a very moderate sized ball of earth would contain all of the roots necessary to the well-being of the plant.

Transplanting should be done as early as possible. November, February and March I deem the best seasons here. Plant no deeper than the tree grew

18. Venetian sumac (*Cotinus coggygria*) is native to Asia; American smoketree, *Cotinus obovatus,* is native to North America but not generally found in the Gulf South. More commonly found in Affleck's region would be the white fringe tree (*Chionanthus virginicus*), also known as grancy gray beard. The plants have similar characteristics: with Venetian sumac, the flowers would be "delicately colored" (haze-like, smoky white to pink), while the fringe tree's flowers are white.

naturally. Mix the manure with the best of the soil that came out of the hole; when the manure is rank and coarse, best put the larger portion near the surface. Finish with a few buckets of water, and stake the tree or plant firmly that it may not be shaken with the wind.

In removing large trees, thin out the top somewhat, and shorten the branches; but never trim the tree to a bare pole, or anything approaching to it. For although in some instances trees thus treated may live, they will be exceptions to the rule. In transplanting Live Oaks, I prefer clipping off a large portion of the leaves, first shortening the branches. I have a very handsome lot of these, now five to six feet high, which have been twice transplanted and root-pruned, so that they may now be removed with entire safely.

In conclusion, let me advise those who have places to improve, to secure good sized plants, if such can be had that can be safely transplanted. A pleasing effect is thus produced, and at once, which would otherwise require long years of waiting for.

THOMAS AFFLECK

# 10

## Letters from Texas

As a correspondent to the Houston *Telegraph,* Affleck regularly contributed articles on seasonal planting, eliminating pests, and other subjects of general interest to farmers and domestic gardeners in Texas. The first letter (October 6, 1858) given here is addressed to fellow horticulturist and journalist Edward Hopkins Cushing (1829–79), or "Friend Cushing," owner and editor of the Houston *Telegraph.*[1] Here Affleck returns to topics about which he had written earlier, such as wine-making, sorghum, and plants appropriate for hedging. This letter later appeared in the Dallas *Herald* (October 27, 1858), suggesting that Affleck's advice held interest for those in Dallas as well as Houston. The second and third letters (August 1, 1859; August 22, 1860) also appeared in the Houston *Telegraph,* and they contain Affleck's impressions as a new resident of Texas and observations about agricultural conditions in the state.

Later letters, not given, ask fellow Texans to supply "horses enough to mount, at least, those now afoot" in the Confederate Army, together with a "few pieces of strong, warm clothing, and a pair of stout shoes" (August 25, 1862); another (September 6, 1864) seeks to dispel a rumor questioning his "loyalty to the South" generated by the report of a letter from his "two sons, at school in New York" who wish to come home. In fact, of Affleck's two sons, one was already serving in the Confederate Army and the other was twelve years old, living at home and "already longing for the time when he may join his brother."[2]

1. E. B. Cushing, "Edward Hopkins Cushing: An Appreciation by His Son," *Southwestern Historical Quarterly* 25, no. 4 (April 1922): 261–73. See also http://www.tshaonline.org/handbook/online/articles/fcu34, accessed July 19, 2012.

2. Isaac Dunbar Affleck (1844–1919) served in Terry's Texas Rangers. Much of his Civil War correspondence has been published. He later became an entomologist. See AFFLECK, ISAAC DUNBAR, Handbook of Texas Online (http://www.tshaonline.org/handbook/online/articles/faf01), Texas State Historical Association, accessed July 20, 2012. Recall that Robert Webb Williams Jr., whose essay on Affleck appears earlier, in the 1960s published articles on the younger Affleck's Civil War correspondence.

A later letter discusses another issue of interest to Affleck and others, that of European emigrant labor, and how it might be used to rejuvenate the postwar Texas economy after the abolition of slavery. A report appeared in *Flake's Bulletin* (June 1, 1866) of Galveston, discussing a convention held the previous week to discuss how Texas might emerge from the "chaos in which we were left by the Confederacy," during which two proposals were presented and discussed: the first, proposed by Affleck, "wanted to procure labor, the second to sell lands." The *Bulletin* endorsed the former: "We took the ground that our only refuge from poverty, and our only road to speedy wealth, lay in the direction of emigration; that by bringing here the industrious population of Europe, and landing them on our shores, free in all the attributes of manhood, that we should speedily develop our resources and become great; that our monster plantations would be purchased and cultivated by small farmers, to the enrichment of their present owners and the general good of the whole community." This scheme had been developed by Affleck and published, according to the *Bulletin*'s account, "in a series of articles published in the Galveston News." The scheme received enthusiastic support from the convention's delegates, and a later convention was proposed for Austin to discuss the idea in greater detail. On June 23, Affleck's earlier letter to the Galveston *News* from April 19 was published in the Dallas *Herald* (given here), in which Affleck, en route to Europe, discusses his plan for the immigration of European workers (particularly those from England and Scotland) into Texas as farm workers. In retrospect, the specifics of Affleck's plan are, on one hand, painfully racist, in that only "white men of the best European races" would be allowed and "coolies or . . . any other low-grade race"[3] would not be involved. But on the other hand, his plan seems feasible in concept: steamships would depart from Liverpool for Galveston with workers from England and Scotland and return with livestock to invigorate English herds recently diminished by the "cattle plague." The workers, their passage having been paid by prospective employers without the expectation of recompense, would then have jobs on Texas farms for which they would be paid by "an interest in the crop, or for a term of years."

3. Chinese laborers came to the South during Reconstruction, having been recruited by planters seeking to fill the vacuum of agricultural workers resulting from emancipation. Efforts by some planters to create a new form of slavery, however, were largely unsuccessful, and the results of Chinese laborers in agricultural situations were mixed. See Lucy M. Cohen, *Chinese in the Post–Civil War South: A People without a History* (1984; Baton Rouge: Louisiana State University Press, 1999).

We see in Affleck's letters, as in his previous writings, an interest in practical matters and topical issues, together with a willingness to share opinions based on his experience and informed discussions with others. And while we may not necessarily find prescient ideas that anticipate future economic, social, or political realities, we do find innovative ideas that address contemporary issues in pragmatic and entrepreneurial ways.

In Edward Hopkins Cushing, Affleck certainly found a kindred spirit, and their friendship was mutually beneficial to the careers of both. "Friend Cushing" grew up on a farm in Vermont, and following graduation from Dartmouth in 1850, sailed to Galveston, where he first taught school, then wrote for a local newspaper. In 1856 he acquired the Houston *Telegraph* and became its editor. At the time, Houston was a small but growing community of about 2,500 residents. Through his editorial columns, Cushing was an active booster for Houston's development, supporting a variety of interests such as education, horticulture and scientific farming, animal husbandry, manufacturing, and railroads, and he encouraged Texas authors by printing their works. Many of the topics that "Friend Cushing" supported coincided with Affleck's interests.

Despite difficulties, Cushing continued to publish his newspaper throughout the Civil War. Although he was a secessionist, once the war was over he advocated for working to address the aftermath of war and occupation, opposing corruption and seeking to expose graft where he saw it among carpetbag politicians. Cushing sold the *Telegraph* in the late 1860s but retained an interest in publishing. He published Affleck's *Hedging and Hedging Plants in the Southern States* in 1869.

—Ed.

## From the *Weekly Telegraph,* Houston

October 6, 1858

*From the Telegraph.*

LETTER FROM THOMAS AFFLECK.

Central Nurseries, Washington Co.

Texas, near Brenham, August 25, 1858

*Friend Cushing.*—I have looked for you here, now and again, ever since we began wire-working, but as yet in vain.

One of your subscribers at Eutaw Limestone Co., propounds enquires [*sic*] enough, in a letter received to-day, to require the contents of a two hundred page 12 mo volume to reply to in full! Others have made like enquiries. With your leave I will answer them through the columns of the Telegraph.

Our Eutaw friend requests me to detail the modus operandi of wine-making from the Mustang grape.[4] The manner of obtaining the juice, the amount of water used, if any, and so on.

Although tolerably conversant with the manner of making wine, as practiced in the different countries of Europe, and about Cincinnati &c., I find that there is much to learn from practice and experience, with a grape like the Mustang, newly employed for the purpose, or, at least, of which we have not written experience. For my part I have been mainly a looker on this year, a pupil of M. Gérard. Until [learning of] his experience in a small way, in 1856, I had strong doubts whether a wine made from the juice of the mustang would keep in this climate without the addition of spirit or of sugar which becomes spirit. The wine made that year, and that too with very ordinary means, such as tubs &c., was excellent, and kept perfectly; being now sound and good; vastly improved indeed by age. Nothing was used but the juice of the grape, expressed by tramping.

4. Eutaw was in Limestone County, in East Texas, formed in the 1840s by settlers from Eutaw, Alabama. In the late 1860s, a new rail line bypassed Eutaw in favor of Kosse, a nearby town, and Eutaw slowly disappeared. The Mustang grape (*Vitis mustangensis*) is a native species of grape found in East Texas, Oklahoma, Louisiana, Arkansas, Mississippi, and Alabama. It produces small clusters of hard green fruit that ripen into thick-skinned, soft dark-purple berries between June and September. Because of its high acid content, the fruit can irritate skin and is bitter and unpleasant to eat. The vines can extend over several hundred feet, covering nearby shrubs, small trees, fences, and other objects. In spite of its high acidity, the fruit makes an acceptable wine when sweetened and diluted with water.

This year the grapes were gathered by plucking off the branches with the finger and thumb. Every second evening, the grapes thus gathered, during the two days, were run through one of W. O. Hickok's cider mills (made at Harrisburg, Pa., and an excellent machine it is for the purpose) juice, skins, pulp and seeds all running directly into a fermenting tub. Fermentation commences immediately; the mass rising to a considerable height, care being taken not to break the crust or mass of skins &c. To say what "the proper temperature should be" or the "length of time required for fermenting" &c., &c., is impossible for me as yet. My cellar is 30 by 15, and ten feet deep, with a stout roof of cedar, covered with earth, and kept *as cool as possible.* The time required for fermenting varies from 60 to 80 hours; and can only be determined *by the nose.* The tub may be tapped with a gimblet[5] and a little wine drawn off. When it runs clear and smells, well, *like wine,* is the best direction, it may be drawn off and barreled. Fill the barrel and lay on the bung reversed, so as not to be tight. After from two to three weeks, the wine may be drawn off clear, and still further clarified with the white of an egg. Then when fully settled, say in two or four weeks, it may be bottled off, or racked off again into casks to remain until wanted.

When the wine has been drawn from the tub, until it begins to appear muddy, stop it; then add three or four buckets of water to each barrel of grapes in the tub, and allow another fermentation, when the result will be *piquette*[6] or *petit wine*—the most excellent drink imaginable during warm weather; and may be drank as soon as made. It will not keep long.

These directions might be greatly extended. But the fact is that those who desire to do more than make a few barrels for their own use had better employ some one in the first instance, who understands fully the process.

I intend planting in vineyard, the mustang grape, every foot of land under tillage this coming fall.

Mustang wine requires age, and repeated racking off and other manipulations, to bring it to anything like the degree of perfection it is capable of being brought to. It is an excellent wine, but in its natural state contains a very large proportion of tartaric acid, which, however, it precipitates in the cask with time. Drs. Key, Graves, Red and other practicing physicians here,

5. Affleck likely means *gimlet* here, a small tool with a screw point used for boring holes.

6. A French term referring to a mild, simple, winelike beverage made by adding water to grape pomace (the pulpy mass of grape skins, stems, and seeds).

prescribe it, whenever to be had, as the best of all tonics to patients in a state of convalescence; and especially after low fevers.

My correspondent enquires farther—"the number of gallons of syrup that the common sugar cane will yield to the acre in Texas, when grown; the number of acres usually cultivated to the hand, and the number of gallons of juice to make one of syrup? I want this information to compare it to the Sorgho."[7]

I will leave you to reply to these queries.

A very considerable quantity of syrup has been made generally, throughout the interior of Texas this season. In this vicinity several mills and furnaces have been erected and much excellent syrup made, and which will serve the general purposes of *sweetening*.

My correspondent's chief topic, however, is *Hedging,* and with the Cherokee Rose.[8] He understands that I propose to "accept of contracts on liberal terms" to hedge for others. On this head let me refer him to your advertising columns. And further permit me to say, I am by no means wedded to this particular plant, the Cherokee Rose, so called; but would be glad indeed to find another better adopted to the purpose. In the *double white microphylla* Rose, I think we have one equally as well, perhaps better suited to the stiff black prairie lands, because equally hardy, well-armed and impervious, yet not so rampant. For rich bottom lands I prefer the so-called *Chickasaw* Rose, which resembles the Cherokee, but with smaller, closer, squatty, evergreen foliage and less rampant habit.[9] Upon the whole, however, the Cherokee may

7. Sorgho, or sorghum, is a grass (*S. vulgare*) with many varieties that produce grain used throughout the world for food. In the American South, sorghum was grown as food for livestock. Its grains can be pressed to make a thick syrup similar to that of sugarcane.

8. *Rosa laevigata.* This rose takes its name from its association with the "Trail of Tears," the forced relocation of the Cherokee, Muscogee/Creek, Seminole, Chickasaw, and Choctaw nations from their ancestral homes in the Southeast to Oklahoma in the 1830s, following President Andrew Jackson's Indian Removal Act of 1830. Many died along the way of disease or exhaustion. The rose is thought to have been introduced into the Southeast in the late eighteenth century, and it soon became naturalized throughout the region. According to legend, its petals represent the tears shed during the period of hardship and grief throughout the march, the gold center symbolizes the gold taken from the Cherokee nation, and the seven parts of each leaf their seven clans.

9. *Rosa bractaeta*, the Chickasaw rose, also known as the McCartney rose, is identical to the Cherokee rose in all respects, except its leaves have seven to nine leaflets instead of the three to five found on the Cherokee rose. It is a vigorous grower in just about any soil, vicious with thorns, and resistant to the diseases that plague other roses. Its leaves are glossy green, and its flowers are strongly fragrant.

prove the safest, if the experience of hundreds running through a period of fifty years or more in South Carolina, Georgia, Mississippi &c., goes for anything, and especially considering the thousands of miles of thorough fence of this plant which exists. That it will suit equally well [to] all soils and localities may well be doubted. For the low sea-coast prairie the *Guisachee* or *Weesachee* will, no doubt, prove to be the better hedging plant.[10] For the thinner and poor upland prairies it is more than probable that the Osage Orange will be the best adapted.[11] I had great confidence in the restive Cock's spur Hawthorn until this season, when it has been almost entirely destroyed by a small insect of the Aphis family, which operates underneath the leaf; and which same insect has been very destructive to the Quince trees. And from dread of this same insect, I say nothing of the *Pyracantha* thorn, though as yet no damage has been done. In Western Texas, the *Yucca,* a Spanish bayonet, and the *Opuntia* or Prickly Pear, either separated or mixed together, in the end could I feel confident, be made to form an impassable barrier. There is a small tree here, of the *Rhameus* or Buckthorn family, called by [the name of] India-rubber tree of which, too, good hedges could be made.[12]

In this, however, it is not a question of what skill and industry could or might do, but of what may be done economically and profitably. The plant which when made into a hedge *must* receive a certain and very considerably expenditure of labor *annually* to keep it in the form and serve the purpose of a fence, is of doubtful value here for that object. That some labor ought to be bestowed on a hedge of any kind, at least once a year, is sure. But that plant will best serve our purposes, which will hold its own through a year of neglect, should circumstances compel its being neglected, and will still continue to form a good fence, and may again be brought into its proper

10. Huisache (*Acacia smallii Islay*) is a small tree or shrub, native to South Texas, of up to thirty feet in height with thorny branches that re-sprout readily when disturbed. This shrubby plant is attractive to wildlife: birds and small mammals nest in its cover, insects are attracted to the nectar in its flowers, and deer eat its fruit. It tolerates dry conditions, and its wood is used for firewood and fence posts.

11. Osage orange, *Maclura pomifera,* is sometimes known as bois d'arc. This tree's hard, grapefruit-sized fruit was thought to repel insects, and its dense, hard wood was used for fence posts and tool handles.

12. A summary of this and the following paragraph was published in the *Annual Report of the American Institute of the City of New York for 1858* (Albany, 1859), 170–71, beginning as follows: "Thomas Affleck of Texas advertises in their papers he will plant and dress for three years hedges of the *Cherokee Rose* for $100 a *mile.* He says the double white *Microphylla* Rose is better."

form as a perfect hedge, as soon as circumstances will permit the bestowal of the labor needful. In the three roses named, I think we have these plants— Cherokee, Chickasaw, and white Microphylla.

It is true, I could undertake contracts to hedge, this fall, at so much per mile. But it is doubtful if any have their hedge-rows ready for the plants. Hence it would be better to have that done thoroughly during the coming winter, and the planting could then be done to some good purpose the following fall. Still where the intended lines of hedge have been already in cultivation, and not less than from two to twenty miles in a neighborhood could be contracted for it could be done now. The terms as named in your advertising columns, are $100 per mile: payable one half when the hedge is planted, one-fourth after the dressing and replanting is done the next fall, and the remaining fourth the fall following. The parties for whom I hedge must prepare the hedge-row thus—It should be thoroughly plowed (having been broken up at least one year before) to the width of ten feet, and as deep as possible. If done this winter to be planted next fall, a crop of cotton, peas, or sweet-potatoes might be taken from it. As early as practicable in the fall it is to be planted, it should again be well broken up and harrowed, ridging to the centre. I then furnish the cuttings of young plants, and plant. The young hedge must then be tended by the owner, as if it was so much cotton. The following fall I prune and dress it, replanting when needed. And again the owner tends as before, with some little additional work as I shall direct. The next fall, I again prune, dress, and arrange it, leaving it an entire unbroken line of young hedge, which if properly treated, will form a perfect fence the fourth year. I plant two rows of cuttings giving a close and solid bottom to the hedge.

I think our Eutaw friend will here find all of his enquiries answered.

How lovely the face of this beautiful country now, and how positively delicious the weather! True we need rain for the pastures in the prairies. In the timber the range is excellent, and the most abundant. Corn is plenty and cheap, ranging from 30 to 50c. Cotton is a short crop to what was expected. One good old neighbor who, with one good soaking rain felt sure of a 500 lb bale per acre—Upland—complains sadly that one half his crop will not yield more than 1,000 lbs. per acre, and the remainder about 800 lbs.! I think he has no very great cause to complain. Some will do better than this, a good many not nearly so well. But for all the labor put upon crops here, I should say the yield is wonderful.

Notwithstanding that I have not had rain enough, this summer, to have kept a plant alive in my nursery, had the soil been as in Mississippi, everything looks green and well. The apple trees begin to drop some leaves; and so do peaches and nectarines. But pear, fig, apricot, plum, &c trees are yet as full of foliage as if we had had weekly showers; and I could show you, and do hope yet to hand you, some pears as fine as ever grew, of late Peaches, too, I see a few on some trees. Recollect, this [is] only my second year. The inferior varieties of European grapes, too, are yet green and vigorous.

But above all, the rose in its varieties, is thrifty and vigorous to an astonishing degree. We want rain for a full autumn bloom; and a little *nitrogenious* [sic] manure—*something that smells pretty strong!*—worked in around them with the first good rains would help the flowers, but is not needed for the growth.

Asparagus is yet as green and *growing* as in spring. Nothing surprises me more than this, and what a delicious vegetable it is!

But I must hold up! You little know the risk you ran when you offered me space in your columns for remarks on Horticulture and Agricultural topics! Being rather apt to let my hobby run away with me.

By the way—In preparing for hedgerows for planting: wherever prairie soil is turned up in which it is known that cotton would rust or die out, a tolerably good dressing of cotton seed or other equally good manure should be applied before planting.

Yours truly,<br>THOMAS AFFLECK.

---

**From the *Weekly Telegraph*, Houston, № 2**

AGRICULTURAL THOUGHTS AS THEY ARISE,
OF A NEW SETTLER IN TEXAS. NO. 1.

Central Nurseries, near Brenham
Washington co., Texas, 1st Aug., '59

ED. TELEGRAPH:—It is wonderful the number of persons who are, at this time, looking towards Texas as a future home, in all of the older slave States.

All desire information about the climate, soils, crops, health, &c. This is a somewhat central region. The climate too, is a medium one; about the dividing region of those portions of the State in which droughts have, of late, prevailed; and the wetter regions of Louisiana and Eastern Texas, and although well to the southward as to latitude, yet the great elevation above the sea, and the high rolling character of most of the surface of the country, give us fresh and cool sea-breezes at almost all times, that carry us far to the north of our actual latitude.

Having only settled down here this spring, many things will perhaps strike me to be worth noting as of interest to the enquirer at a distance, that would escape the older residents.

Hence I have thought that a few notes embracing such matters, ranging through the week, and from week to week, will be read with interest.

I cannot promise great regularity, either in communicating with you every week, nor in noting the weather, range of the thermometer, &c. But will be as exact as circumstances will admit of.

Nearly all of this vast State has suffered, for three years past, from the severest droughts ever experienced, within the knowledge of the oldest settlers. Yet trying as the drought has been, it would have done but little damage any of these years, but for the extraordinary late Spring frosts, which have cut down the crops at periods so late as to render a crop from the replant very uncertain. Just think of it—a killing frost, this last Spring, on the night of the 22, April! And yet, thanks to the grateful rains which followed, and the wonderfully generous soil of this great State, the heaviest crops of wheat and other small grains, and of corn have been made, and the best prospect for a full crop of cotton nearly all over the State, that has perhaps ever been known.

In some portions of the west and northwest, the crops generally have been an almost total failure during these three years past. And yet nearly all have been able to maintain themselves, through the great increase the thriftiness of their livestock, and few can be found willing to leave that lovely and perfectly healthy country for any other, notwithstanding all their trials.

Through the central regions of the State, somewhat short crops were made. Yet it has been wonderful to note the effect of deep and thorough breaking up of the ground, early planting and prompt replanting when unfortunately necessary, and clean and thorough tillage, when compared with the opposite course, perhaps just over the fence.[13]

13. Deep plowing was a technique advanced through scientific farming in which fields were plowed to a depth of around twelve inches, breaking up the soil, bringing nutrients to the surface, overturning residue from previous crops, and preparing the ground for new plantings. In general, this technique resulted in increased yields.

These black prairie lands, these of the river and creek bottoms, and the better class of pasture lands, produce good crops with less rain than any other that I have any knowledge of.

We have had seasonable rains during all of this Spring and Summer in this district. They were light, and have not saturated the ground to a depth sufficient to replace the moisture dried up during these last three years. On digging a well, I find the ground moist to the depth of three or four feet, but dry as powder for the next fifteen or twenty. The consequence is that all springs which do not originate at a great depth have more or less failed. Still we have water in sufficient abundance.

Our last showers fell here about two weeks ago—fair summer showers. Corn is made, and fodder pulled. Cotton will soon need rain. The health of the country is good. Newcomers, during their first summer, and having no cistern water, may expect an occasional chill, but of a light nature, easily broken up. But almost sure to return on the 7th 14th and 21st day after the last chill, if not guarded against by a few grains of quinine, and perhaps a blue pill the previous days. I have heard of no sickness of a more serious nature.[14] The weather is hot in the sun-shine from eleven until three o'clock; but while a fine breeze [is] blowing almost all the while. Thermometer, to-day, at noon, in the shade, 88.

What a comfort and luxury are these cool evenings and nights, without the hum of a single mosquito! However hot and exhausting the days, the wearied laborer is sure of a sound and refreshing night's sleep.

Who has ever made an exact calculation of the cost of the fodder crop? The time and toil in the close atmosphere between the corn rows during this hot weather, whilst the cotton suffers for a less thorough working, before the business of picking begins. None of us sow half enough of winter oats, millet, &c.

And how much additional labor is lost by the lack of a fodder-rack to the wagon, and of a sufficiency of steel forks to pitch with, and the want of a large stable or barn; with a loft overhead roomy enough to house all the rough feed, as fodder, millet, &c. the getting, trimming and planting roles, and cutting brush as a foundation for fodder stacks is another item of time lost.

14. Affleck here is likely referring to a slight sickness—"an occasional chill"—that might come to new residents unacquainted with local water, the remedy for which would be quinine, a natural white crystalline alkaloid having fever-reducing, antimalarial, pain-killing, and anti-inflammatory properties. It occurs naturally in the bark of the cinchona tree, indigenous to South America.

The new-comer to a new country must include as a part of the cost of his fine body of rich land, at a comparatively low first price, the loss of all the little conveniences he has been all of his previous farming life, accumulating around him.[15] Each is itself of a triffle; not so the aggregate.

Young stock require occasional looking after at this season of the year. . . . [*Here the existing newspaper is undecipherable; three paragraphs follow, all concerned with livestock management.—Ed.*]

And yet a flock of very fine Saxony sheep, driven through to this vicinity this Spring, from the Mississippi, are as quiet and gentle as ever. My jolly big long wooled sheep take [indistinct]: But they fatten almost too kindly, so much so, that I have to starve them a little—keeping them in the pen til 10 o'clock each day.

Yours,

THOMAS AFFLECK

————◆————

**From the *Weekly Telegraph,* Houston, № 3**

Letters from Thomas Affleck.
Washington County, August 22, '60

*Editor of the Telegraph:*

A glorious, dense, soaking rain has been falling here since yesterday at noon; worth a million to this region. If the fall is not an *unfavorable* one, the cotton crop of this county will be doubled: and, if the planters and farmers act with any degree of prudence and common sense, in putting in every acre they possibly can, of winter small grain, the stock will get through the winter without serious loss.

I am getting in turnips, beets, carrots, and every other vegetable that *may* succeed: I have several acres of drilled corn, now from 4 to 6 inches high, since the rain of last week: and I hope to sow ten acres more to-morrow, if it ceases raining, or when it does. I *hope* to make a fine lot of fodder. I *may* fail, but *if I don't sow the seed,* I cannot possibly get the fodder.

By the way, if I might suggest, with the limited experience I have of the

---

15. Affleck suggests here that the relatively cheap price of good Texas land is worth the "loss of all the little conveniences" previously enjoyed elsewhere.

soil and climate of this portion of Texas, I would say, so soon as the ground is dry enough—stir it between the rows of cotton.

Small grain may be sown to advantage in this way. The after treading of the ground during the cotton picking, will help the grain crop.

Yours,

T. A.

---

## From the Dallas *Herald,* June 23, 1866

Letter from Mr. Affleck.
From the Galveston News

On Board Steamer Persia
Within 100 miles of Sandy Hook,
April 19, 1866

Eds. News.—I find in my portfolio several letters commenced for you; but which it was out of my power to finish and send off. I will now give you an outline of what suggests itself which may interest your readers.

You know what I aimed at in taking this trip. I am glad to say that I have been successful beyond my most sanguine hopes; although a great deal yet remains to be done to produce the wished-for results to the State.

We want laborers and population of all productive classes. And, to make Texas what she is fully capable of becoming, these immigrants must be white men of the best European races. Let nothing tempt our people to encourage the introduction of coolies or of any other low-grade race. To import this needed population shipping is required, and that too, propelled by steam. Steamships are pretty costly affairs to run. No man will risk them unless the results are made very clear. This can only be done by an assurance of full and paying loads each way; or by a State subsidy in the beginning of the enterprise.

Steamships, to be profitable, under any circumstances, and to cross the ocean speedily and in safety, must be of good size—too large to cross the bars of our harbors.[16] To ride outside at anchor, lightering the cargoes both in and

---

16. Here, the term *bars* refers to sandbars. These fluctuating beds of sediment prevented ships of certain size from entering Galveston's harbor. Affleck, at the end of his life, was an ac-

out, is an expensive item, to be paid by the shipper and consumer. To avoid this and insure regular lines of good ships, a ship channel must be dredged through the bar.

The merchants of the State must aid in the steamship enterprise, by pledging their freight, both out and in, to the line. If once established, one line proving a success, would quickly lead to competition, with its natural results.

I learnt that an enterprise, which I had long thought of as being perfectly practicable—the exportation of livestock, as cattle sheep, horses and mules, from Texas to England—was proposed in Liverpool, and, judging from a prospectus of the company proposing it, was actually about to be carried out; and thinking that, in this there was a prospect of direct and cheap transportation for emigrants, to our State, I hastened on to do all in my power to foster and assist the enterprise, and also to satisfy myself and the people of Texas what the prospect really was for turning a tide of European emigration to our shores. I found that the "English and American Cattle Traffic and Transit Company," which I had been led to believe was fully organized and under way, was only proposed. Its projector, Mr. W. A. Hayman, of Liverpool, sought to *float* the company, as it is termed; and had engaged the attention of a number of able business men in that city. They thought so well of the enterprise, in view of the great destruction of cattle from the dreadful Rinderpest, or cattle plague,[17] that they determined upon sending out [a] screw steamship, to be fitted up for the purpose, and test it by the importation of one cargo of fat beef cattle. I and others gave evidence before the Board of Enquiry, as to the probable result, etc., more than once. But there proved to be a kink somewhere. I waited the unraveling of it until my patience was exhausted, then took the matter in hand myself; and believe that it will be carried out after the completion of some necessary preliminaries.

The greatest difficulty in the way is the condition of the entrance to Galveston Harbor. That is an impediment which must be, as it speedily can be, removed.

---

tive participant in efforts to dredge this harbor to make it easier for large ships to enter, thereby increasing economic activity in the Galveston/Houston area. These efforts failed, investors lost their money, and controversy ensued.

17. Rinderpest is an infectious viral disease of cattle. The term *Rinderpest* is German, meaning "cattle-plague." In 2011, the United Nations declared that it had been eradicated, making it only the second disease, after smallpox, to be fully wiped out.

I bring with me assurances from Liverpool shipping merchants, that a line of screw steamers will be established to run once a fortnight, between Liverpool and Galveston; the first ship to sail early in September, on conditions which can be carried out, I believe.

The transportation being secured, there will be no difficulty in inducing any amount of emigration, of the best class of English, Scotch and other laborers to Texas.

Immediately on landing in England, I wrote, printed and circulated a little twelve page pamphlet, briefly descriptive of Texas, and explanatory of her wants in the way of population. These I scattered over the length and breadth of the land—no trifling job, let me tell you, and costing something besides. I wrote many letters to the principal papers; which were generally printed; pointing out the varied resources of our State, and the inducements these offered both to immigrants and to capital. Calling attention to the immense supplies of beef and mutton which might be easily procured from us, to supply the growing wants of the people of Great Britain; and showing the different ways in which that meat could be safely and surely carried across the ocean. The results were gratifying in the extreme. My whole time thereafter was taken up in answering hundreds of letters of enquiry upon, it seemed to me, every imaginable point, and in keeping appointments at different private places, both in Scotland and England, to talk with the people, exhibit maps, explain the climate, soil, resources, &c., of Texas; and to give generally all the information in my power. I found an anxious desire to benefit by the prospects held out, of advantages to be gained by emigrating to these States of the late Confederacy, and more especially to Texas. I found a most friendly and sympathetic feeling amongst the people, rich and poor, in our favor; and which prevailed universally during our struggle, and from the very first.

I found a wish to invest money in any way that afforded a clear business prospect of fair returns—in banking, lending to planters, to enable them to renew their planting operations, extending railroads, assisting emigration, sustaining steam, communication, curing, packing and importing meats and livestock, &c. &c.

It is to be hoped that the Convention, which I presume is still in session, will do all in their power to sustain, and nothing whatever to damage, the credit and good of the State. Everything depends upon that, as I shall show.

I found that any number of laborers, house servants, mechanics, skilled operatives, clerks, salesmen, teachers, &c., would gladly come out, and engage themselves on fair terms, if the means and the pay were made clear to them. A hundred thousand or more of farm laborers to work on the plan I proposed, for an interest in the crop, or for a term of years, can be procured, if a proper course if pursued.

As quickly as in my power, I will enter into particulars in my letters to you.

THOMAS AFFLECK.

Of Brenham, Washington Co., Tex.

P.S. As it is impossible for me to write more than one series of letters, and as my time and services were given, and a long journey taken with a view of benefiting Texas, I beg of the different papers in the State that they republish these letters, and send me copies of the papers containing them.

# 11

From *Hedging and Hedging Plants,*
*in the Southern States* (1869)

Affleck envisioned a series of books of a horticultural nature intended for
the southern audience, but only one appeared, and it was published posthumously. Its subject—hedges and plants that might be used for hedges—was
one he thought was pertinent to landowners in the South and a subject about
which he had written earlier.

Included here are the book's preface and introduction, in which he discusses the relevance of his subject. In the remainder of the book, he discusses practical matters such as plant propagation, water required, soil
characteristics, effects of weather, hedges for screens and fences, and the
preparation of the hedgerow, leading to a discussion of what he thought were
the best plants for hedging in the South.

Affleck mentions that discussion of the Cherokee rose as a fence had
already appeared in several agricultural newspapers as contributions of
correspondents from the South. He notes that in 1832 he begin investigating hedges of English hawthorn (*Crataegus oxycanthus*), cockspur (*C. crus-
galli*), honey-locust (*Gleditschia triancanthos*), and osage orange (*Maclura
pomifera*) in Delaware, Pennsylvania, Indiana, and Illinois. For the most
part, these were unsatisfactory until "I visited Mississippi, and saw those of
the Cherokee and Chickasaw (*R. bracteata?*) rose." After he settled in Mississippi in 1842, he "looked more closely into the matter of hedges, gathering
all that was possible from the experiences of others as well as from close
observation and active practice." From this effort, he continues, it "was impossible to reach any other conclusion than that, in these and others of the
rose family, we had precisely what was wanted for Southern plantation and
farm fences." Affleck notes that he has "repeated attention to it through the
press" (a chapter appears in *Southern Rural Almanac,* 1847, and "a more
elaborate" treatment in *De Bow's Review,* January, 1848), and through his

efforts, "between 1850 and 1855, hundreds, if not thousands of miles were planted."[1]

Then follows a lengthy discussion of the Cherokee (and Chickasaw) rose, covering its beauty; strength; permanency; resistance to cold weather; length of time required to make a fence; objections ("the great space occupied by an unpruned Cherokee Rose hedge is almost the only objection made"); propagation; and method of planting. He briefly discusses other roses as well.

He includes a discussion of the osage orange, partly because so much had been made of its use throughout the country as a hedge. Affleck notes that it will form "a *thorough fence*, but never a *neat and perfect hedge*, such as a hedge ought to be." Nevertheless, he listed its advantages, including its widespread adaptability to a variety of conditions, together with its cheapness, durability, and longevity. Affleck gives instructions for planting, cultivation, and pruning into a hedge form, but this requires labor, and Affleck wonders if this plant is worth the trouble it requires to shape it into a hedge, when other options, such as the Cherokee rose, pyracantha, or Weestachie (*Vachellia farnesiana*)[2] grow and thrive.

Next are short discussions of other plants, such as cockspur (*Cratægus crus-galli*), prickly pear or cactus (*Opuntia*, var.), American aloe (*Agave Americana*), and Spanish bayonet (*Yucca gloriosa*). Finally, he gives brief suggestions and propagation instructions for ornamental hedges. Affleck's conclusion gives four "deductions" for consideration when selecting a hedging plant: first, the plant selected "*must be*" regularly pruned; second, it should be easily propagated, must thrive in a variety of conditions, not sucker, and not exhaust the soil; third, rather than a "vigorous and persistently upright growth," the plant should be of a "partly trailing and pliant or horizontal habit"; and finally, it should be evergreen. Finally, for hedging above the "Southern line of North Carolina and Tennessee," he recommends the *Alba odorata* (Maccartney) rose, the evergreen hawthorn, and the *Cratægus pyracantha*. For areas south of that line, "take the Cherokee rose" is his advice for "not very rich land." Other suggestions include the Chickasaw rose for better soil, the Maccartney rose for rich bottom land, and the

---

1. Recall that planting such hedges was a service that Affleck would provide for a fee.

2. *Vachellia farnesiana,* also known as *Acacia farnesiana,* commonly known as needle bush, is so named because of the numerous thorns distributed along its branches. The native range of *V. farnesiana* is uncertain. This may—or may not—be the plant mentioned in note 10, page 225.

pyracantha. For a fence near a homestead, "I would trust the Manettii rose," and for lowlands near the coast in the West, "I would rely upon the Weesatchie (*Vachellia farnnesiana*)." As far as the osage orange is concerned, he advises that it be used with caution; while it can be a "very beautiful object, whilst in leaf," Affleck says that a "piece of the very best land on my place, in which a few plants were grown twelve years ago, in nursery-row, is as though poisoned and indestructibly occupied by it, as it would be if set in bitter coco-grass (*Cyperus hydra*)."

In *Hedging and Hedging Plants,* we see characteristics of Affleck's earlier works: brief, focused, specific, and easy-to-read. This work is the final statement in a continuum rather than a comprehensive summary of a varied career. And this is one reason why Affleck's writings bear a second look: in his career there is no *magnum opus,* either printed (such as Downing's *Theory and Practice of Landscape Gardening*) or built (as with Olmsted's parks and park systems). Instead, Affleck's career is composed of numerous small contributions on varied subjects, each of minor significance individually, but when taken collectively, they define a career of influence based on curiosity, intelligence, and a genuine interest in helping others.

It is curious that after almost three decades of editing and writing articles on multiple subjects, creating almanacs, and voluminous correspondence, Affleck's career would end, posthumously, with the publication of a slim volume devoted to the seemingly minor subject of hedging and hedging plants. Yet from another perspective, this work closes the circle of Affleck's career, linking his early observations with his final words. What Affleck presents in this work, and what constitutes about a quarter of its content, is an argument for using the Cherokee rose (its genetically close relative, the Chickasaw rose, or other roses) as an efficient way to define property, based on economics, horticulture, and function. As he states in his introduction, he first saw such a fence in south Mississippi in 1841, and he concluded, "it must be universally adopted wherever it would thrive." One reason given earlier for its use in marking plantation property was to keep workers from leaving; now, the reason Affleck gives for the plant's use is based on an efficient way to repair damage wrought by marauding Union troops.

We know he envisioned a multivolume series of numerous horticultural topics, but why begin the series with the subject of hedging? Perhaps it was because in February 1861, while moving his nursery from Adams County, Mississippi, to Washington County, Texas, his entire stock was destroyed when the riverboat *Charmer,* carrying 3,800 bales of cotton as well as Affleck's nursery, burned "to the water's edge" and sank ten miles below Donaldsonville, about halfway between Baton Rouge and New Orleans. Five lives, together with all the cargo, were lost, as was Affleck's entire stock of nursery plants. The results of years of growing and experimenting with plants were gone, and he would have to start his nursery in Texas anew. This was, as well, at the onset of the Civil War, a major reason for Affleck's leaving Mississippi and moving to Texas. As he notes in his introductory remarks to *Hedging,* "During the past eight years all progress has been checked; nay arrested. And many years must roll round before any great change for the better can be hoped for. Yet the effort must be made. And *Hedging,* especially in Texas, seems to me the improvement which must first be attempted."

Another reason might be that, following the Civil War, Affleck was occupied with a scheme that would send Texas cattle to England and bring English laborers back to work on Texas farms. He spent time abroad and lobbied many influential people for their support of this proposal through his writings and personal correspondence. Yet it ended in failure, with charges and countercharges concerning investors' money, all of which was lost. Perhaps Affleck, wanting to continue to exert a degree of the authority in post–Civil War Texas that he had enjoyed before the war, retrieved something already reasonably complete, polished it, and sent it off to his friend, journalist and amateur horticulturist E. H. Cushing, who had by then sold his interest in the Houston *Telegraph,* started a publishing business, and sought to publish the works of Texas authors. If so, the appearance of this work, like many of Affleck's previous ventures, owes more to a convergence of circumstance and convenience than to an orchestrated, well-considered strategy.

As far as American landscape architecture and garden history are concerned, Thomas Affleck may not command a place commensurate with that of Downing or Olmsted, yet the influence of his career is of importance in gaining insight into general environmental attitudes in mid-nineteenth-century America and of specific application to understanding the Deep South region in which he lived. Affleck represents many nineteenth-century figures elsewhere throughout America, now unknown, who made similar contributions to the development of environmental attitudes, designed landscapes, and advocated for practices of land stewardship of the regions in which they lived.

We have volumes about the lives, careers, and influences of those of national significance, yet we know little about these men and women of regional importance. A full understanding of America's environmental history, however, will come only when those of regional significance, such as Thomas Affleck, have been reevaluated and their contributions reexamined through an understanding of the context of their times and an appreciation of their influence on subsequent generations.

—Ed.

## From *Hedging and Hedging Plants, in the Southern States*
### (Houston, 1869)

#### PREFACE

I offer this brief treatise upon Hedging and hedging Plants, to the People of Texas and the other Southern States, without seeing any occasion for apology.

And, believing that I fully understand the subjects treated of, in theory and in practice, and having long seen and urged their importance, deprecate no criticism.

Commonism,[3] in any form, is disliked by the people of the South. No system, which would do away with fencing, could be made acceptable. Every man desires to feel that what he possesses is his own; and is inclined to resist all outside interference. And now, more than ever, is this inclination felt. The "ten-rail fence, double ridered," [a split-rail fence] was formerly looked upon as sufficient; now, even that affords but indifferent protection.

Moreover, fencing materials are becoming scarce. Over great parts of Texas, they cost many times the original value of the land to be enclosed. In the other States, wherever armies passed, the fences were destroyed and cannot now be replaced.[4]

As I shall show, we possess the means of Hedging with plants, which will, one or other of them, grow and thrive in every soil and situation, and form impassable barriers, without the necessity of other protection that certain very primitive means, generally to be easily procured, will afford, and which will be pointed out.

Glenblythe, Washington Co., Texas
1st September, 1868.

#### INTRODUCTION

Fencing by means of Hedges is no new thing. All the world over, wherever plants are cultivated, live-fences are to be found. And a great variety of plants are employed for the purpose.

In these United States, comparatively little has been done in this way, ex-

---

3. A curious word perhaps, but from its context we gather the meaning to be common ownership, in this case applied to land.

4. This situation, a consequence of having been invaded by Northern troops, would have been common throughout the South.

cept in the South. And there is not one mile of thorough live-fence, for one hundred there should be.

Treatise after treatise has been written and published upon the waste of timber in the common rail-fence, where timber is yet to be had for that purpose; elaborate estimates made of the yearly cost of such fences, in time, labor, and material; and endless suggestions made, as to the best substitute to be employed in different parts of the country.

Dry stone-walls have been proposed—that which there is no better fence, if well made, where the material is at hand. Ditches and sod-walls—mere troublesome traps. Stakes and wire—costly and dangerous. Patented plans of wooden and other fences—the most of them mere humbugs. The substantial post-and-rail of Cypress (*Taxiodium distichem*), one of the very best of fences, when well made. As may also be said of the post-and-plant fence, when thoroughly put up, and of good materials. Living posts of china-trees (*Melia Azederach*) or of Black Locust (*Acacia pseudo-acacia*), with nine or ten feet rails of Cypress or other durable timber let in between.

Last of all, it has been by some earnestly and ably urged that fences should be done away with entirely; every farmer housing and feeding his stock, or herding them on his pastures, as in France, Belgium, etc., and in the Connecticut Valley.

All of these modes of fencing have I seen and examined into; and have tested not a few of them.

I have seen hedges in England and elsewhere of Hawthorn, Holly, Crab-apple, Beech, Whin or Gorse (*Ulex europea*), Yew, etc.; screens and ornamental hedges, North and South, of Norway Spruce (*Abies excels*), Hemlock (*Abies Canadensis*), Red Cedar (*Juniperus Virginiana*), Laurier Amandier or Wild Peach (*Cerasus Caroliniana*), which has no equal for this purpose; Yaupon (*Ilex vomitaria*), Cape Jessamine (*Gardenia florida*), Sour and Bitter Orange, numerous Roses, etc.; thorough fences, far to the south, of American Aloe (*Agave Americana*), Prickly Pear (*Opuntia in var*): and others of the Cacti, Spanish Bayonet (*Yucca gloriosa*), etc.; good but costly hedges in the North and West, of several hawthorns, Privets (*Ligustrums*), honey Locusts, etc., and above all, of the Osage Orange or Bois d'Arc (*Maclura aurantiaca*).

But in no country have I seen a fence of any kind, so admirably adapted to the climate and existing state of things, so cheaply obtained and easily kept in order, so permanent, efficient, and substantial, or which has been so thoroughly tested, as the hedge of CHEROKEE ROSE.

I first saw this fence in 1841, in Southern Mississippi; and was so greatly attracted by its beauty and value, I at once concluded that it must be universally adopted wherever it would thrive. But I found, that although many years had elapsed since this plant was first employed for hedging purposes in South Carolina and Georgia, and in Adams and Wilkinson Counties, Mississippi, and that excellent examples of thorough fences of it existed there for a sufficient length of time to have led to its universal use; yet, with even these examples before them, few in those very districts thought of hedging, although timber had become very scarce: and elsewhere the plants seemed to be comparatively unknown.

### A DIGRESSION

This is by no means a rare instance in the history of Southern agriculture; nor, indeed, of the agriculture of the world. Farmers are proverbially slow in adopting improvements; some from indifference, others from a professed dislike for that they call "theoretical" and "book-farming,"—terms freely used when anything out of the old track is proposed; and by not a few from ignorance.[5]

It has been truly said, that the apothegm advanced in the days of Ignorance, and maintained by her children ever since—"To plant well is better than to theorize well"—has been "an instrument of more mischief than any two-edged sword," and of incalculable disadvantage to the agriculture of the South. "Modest merit too often shrinks before it. Let it be asserted, and asserted without the fear of contradiction" (says the same writer—J. E. Jenkins, in the *Southern Agriculturist,* vol. 7, p. 174), "that *theory is the incipiency of all acts;* that the first clod of earth that was ever designedly broken for the introduction of seed with the intention of reaping its production was the effect of speculation and of mental arrangement. To be able to speculate, proves a scrutinizing faculty; and to theorize to success, the highest mental endowment." He remarks, further: "It is easy for the most ignorant clod-hopper to call himself *a planter, and no theorist*—as if he thereby conferred upon himself some honorable distinction, whilst he heaped upon the head of his neighbor—whose mind elucidated his practice, and who is not unwilling that the world should share with him whatsoever of good he can impart—coals of fire and molten lead."

5. Here again, we see Affleck's reiteration of a complaint he voiced throughout his career to those who opposed his recommendations for agricultural reform.

Strong language, this; and yet how true! No improvement is proposed, no new thing introduced to the farming-world, which does not meet with severe checks from self-sufficient ignorance. It is on record, that the cultivation of cotton as a staple crop, and those who sought to introduce it, were included in the same sneering remark—"A fit crop for a petty farmer, but not for a planter."

*Horizontal ploughing,*[6] and the *Side-hill ditch* or *Guard-drain* are valuable instances of "theorizing to success," and require a mind of but very moderate caliber to comprehend their advantages. Yet the first was adopted slowly, and not until the entire State had been almost ruined by the other method of ploughing up and down hill; and is even now unknown or unpractised [*sic*] in many districts. The latter—the *Guard-drain* is as yet a thing but little understood by a great majority of the cultivators of hill-lands in the South. Yet this most perfect preventative to the washing away of the surface-soil, so exceedingly friable on these Southern hill-lands, and the formation of gulleys, has been long known and successfully practiced; and full explanations of the principles upon which they should be laid out and made, and directions for doing the work have been published many years ago, and were very widely made known by myself, through the *Southern Rural Almanac* for 1851. And thus it is with many other improvements of equal or even greater value.

Thus it has been with HEDGING; and not in the South only, but in all the other States as well.

Although I feel safe in saying that the number of miles of Hedge in the South exceeds many times that in all of the other States, yet the progress in Hedging previous to the war was by no means what it should have been. There was a too exclusive use of the one plant, the *Cherokee Rose,* and of its near connection, the so-called *Chickasaw Rose,*—both names being decided misnomers; there being many situations and soils to which neither was well adapted. And where any other plant was tried, it was either the Osage Orange, or some other but little better suited to our necessities.

During the past eight years all progress has been checked; nay, arrested. And many years must roll round before any great change for the better can

6. Horizontal plowing involves plowing with the contour of the earth, enabling topsoil and its nutrients to remain in place rather than washing away, the result of plowing against the grade. Advocated by many, including Thomas Jefferson, agricultural reformers, and editors of the agricultural press, horizontal plowing was not routinely practiced in spite of the inevitable long-term damage of soil erosion caused by other methods of plowing.

be hoped for. Yet the effort must be made. And *Hedging,* especially in Texas, seems to me the improvement which must first be attempted.

I removed from Mississippi to Texas, fully alive to the necessity for this work; and brought with me an immense stock of all kinds, and especially of Hedging-plants and Coniferous trees,—intending that the Central Texas Nurseries should far exceed, in thoroughness and extent, the late Southern Nurseries at Washington, Adams County, Mississippi,—which, I may safely claim, were the first Commercial Nurseries of any great extent in the South.

But it is all said, "L'homme propose, mais Dieu dispose."[7] All of this stock, which had cost many years of toil and expenditure to accumulate was burnt before my eyes, on the unfortunate Steamer "Charmer" in February, 1861. Since that time, during the war and since the surrender, I was able to do little more than test the different plants I thought best suited to the varied wants to Texan soils and localities. And now, in spite of the loss of labor, am prepared to put that experience to a general and practical use.

Each of these plants will be discussed in detail, whether as especially applicable to Texas or to the other Southern States; with full instructions for their propagation and cultivation; the preparation of the intended hedge-row, and the planting and aftercare of the hedge.

7. Roughly translated, "Man proposes, God decides."

# CHRONOLOGY

*Thomas Affleck's Life and Related People and Events*

1812–1832  Thomas Affleck born (1812) in Dumfries, Scotland, to merchant Thomas Affleck and his wife, Mary Hannay Affleck.

Attends animal husbandry/agriculture lectures at the University of Edinburgh; works in the National Bank of Scotland, Dumfries.

1815  Andrew Jackson Downing born in Newburgh, New York.

1822  Frederick Law Olmsted born in Hartford, Connecticut.

1822  The *New England Farmer* (Boston, Massachusetts) appears, one of the earliest agricultural papers published in the United States, Thomas G. Fessenden, editor.

1828  The *Southern Agriculturist* (Charleston, South Carolina) appears.

1831  The *Genesee Farmer* (Rochester, New York) begins publication by Luther Tucker, influential agricultural journalist and editor; becomes a model for similar publications throughout the United States and its territories.

1832  Affleck leaves his position in the bank and emigrates to America, seeking new opportunities.

1832–1840  Affleck travels and works in New York, Pennsylvania, Indiana, and Ohio as clerk/merchant; upon recurring sickness and the death of his wife and child, moves to Cincinnati to recuperate in preparation for returning to Scotland; is financially broken because of the Panic of 1837.

1834  *The Cultivator* appears, Albany, New York, founded by Judge Jesse Buel, an influential agricultural reformer; later taken over by Luther Tucker, 1839.

1838–1848  Olmsted pursues numerous interests and avocations, including seaman, merchant, journalist; attends classes on "scientific farming" at Yale University.

1840  While recovering in Cincinnati from illness, Affleck begins to garden and accepts an editorial position at the *Western Farmer and Gardener,* and, as junior editor, writes for the *Western Farmer* and travels in the Ohio River Valley region, reporting on what he sees.

245

1841    Affleck becomes sole editor of the *Western Farmer;* edits the *Western Farmer and Gardener's Almanac for 1842* (Cincinnati); publishes *Bee-Breeding in the West* (Cincinnati); travels extensively down the Mississippi River to Baton Rouge, Vicksburg, and Natchez on a writing assignment for the *Western Farmer,* sending reports back to Cincinnati for publication; meets his future wife, the widow Anna Dunbar Smith of Natchez.

1841    Downing's *Treatise on the Theory and Practice of Landscape Gardening* appears and becomes instantly popular; Downing is an advocate of Gothic Revival architecture and decoration.

1842    Affleck leaves Cincinnati, moves to Washington, Mississippi, and marries Anna Dunbar Smith (April); assumes responsibility for Ingleside, her failing plantation; creates Southern Nurseries, one of the first commercial nurseries in the South, at Ingleside; conducts experiments on plants, cultivates fruit and plants for southern markets, and imports plants from Europe and elsewhere in America; establishes regional, national, and international business relationships in New Orleans with seedsmen, nurserymen, publishers, and other merchants.

1842    The *American Agriculturist* begins publication, A. B. Allen, editor.

1843    Affleck contributes to the *American Agriculturist's 1844 Almanac.*

1845    Downing's *Fruits and Fruit Trees of America* appears; is instantly popular.

1846    Affleck edits *Norman's Southern Agricultural Almanac for 1847* (New Orleans); continues until the name changes to *Affleck's Southern Rural Almanac* (1850).

1846    Luther Tucker publishes Downing's *The Horticulturist,* a monthly periodical; widely circulated.

1847    Affleck's *The Cotton Plantation Record and Account Book* and *The Sugar Plantation Record and Account Book* are published by B. F. Norman, New Orleans, in editions based on the number of "hands" working the plantation; publication expands and continues at least through the 1859 editions (three versions for cotton plantations, two for sugar plantations).

1848    Olmsted acquires a farm on Staten Island, New York, and initiates the "scientific farming" practices he learned at Yale.

1848    In an editorial in *The Horticulturist,* Downing calls for a large urban park for New York City; local momentum for such a park builds among urban reformers, business leaders, and intellectuals.

1848    Affleck's writings from the *Western Farmer* are published in book form (first printing).

1850    *Western Farmer* articles appear in book form (second printing).

1850    *Affleck's Southern Rural Almanac and Plantation and Garden Calendar for 1851* appears (New Orleans).

1850    Downing publishes *Architecture of Country Houses;* goes to Europe and returns with architect Calvert Vaux, later Olmsted's partner in competition for the design of Central Park.

1850    Olmsted travels to England on a walking tour, visiting Paxton's Birkenhead Park near Liverpool.

1851    Downing introduces Olmsted to architect Calvert Vaux.

1851    Downing publishes Olmsted's article in *The Horticulturist* on Birkenhead Park as an urban example of public open space; is appointed by President Millard Fillmore to design grounds for the White House, Smithsonian Institution, and U.S. Capitol in Washington, D.C.

1851    *Affleck's Southern Rural Almanac and Plantation and Garden Calendar for 1852* (New Orleans) is published; he begins to write regularly for the New Orleans *Picayune* as agricultural editor.

1852    *Affleck's Southern Rural Almanac and Plantation and Garden Calendar for 1853* (New Orleans) is published.

1852–57 Olmsted travels in the American South for the New York *Daily Times,* sending back accounts of what he encounters.

1852    Olmsted's *Walks and Talks of an American Farmer in England* appears; is widely read.

1852    Downing drowns in a steamboat accident on the Hudson River.

1853    *Affleck's Southern Rural Almanac and Plantation and Garden Calendar for 1854* (New Orleans) is published.

1854–55 *Affleck's Southern Rural Almanac*s for 1851–54 appear together in one volume (New Orleans).

1855    *Affleck's Southern Rural Almanac and Plantation and Garden Calendar for 1856* (New Orleans) is published.

1855    Competition is announced for the design of a large urban open space in New York City.

1856    Olmsted's *A Journey in the Seaboard Slave States* is published; is widely circulated.

1857    Affleck moves to Washington County (Texas), near Brenham, and establishes Glenblythe Plantation consisting of a large residential structure together with mills for lumber, cotton, corn, and sorghum and a plant nursery. Affleck moves nursery stock from his Southern Nurseries in Mississippi but everything is lost in transit; Affleck continues work to improve livestock herds in Texas.

| | |
|---|---|
| 1857 | Olmsted's *A Journey through Texas* is published; is widely circulated. |
| 1857 | Olmsted, at the invitation of Vaux, begins work on a design to submit to the competition for a large public open space in New York City. |
| 1858 | Vaux and Olmsted's submission, "Greensward," is chosen as winner of competition for a large urban park in New York City. |
| 1858 | Affleck designs plantings for Louisiana's state capitol in Baton Rouge; plans have not survived. |
| 1859 | *Affleck's Southern Rural Almanac and Plantation and Garden Calendar for 1860* continues, issued from Washington County, Texas. |
| 1860 | Olmsted's *A Journey in the Back Country in the Winter of 1853–54* is published. |
| 1861 | The steamboat carrying Affleck's nursery stock from Adams County, Mississippi, to Washington County, Texas, burns "to the water's edge" with the loss of five lives and all cargo; Affleck must start his nursery in Texas without stock from Mississippi. |
| 1861 | Olmsted's *Cotton Kingdom,* a compilation of previous works, is published. |
| 1861 | Fort Sumpter, South Carolina, is attacked by Confederate troops, beginning of Civil War (April). |
| 1861 | Olmsted begins to work for the U.S. Sanitary Commission. |
| 1863 | Olmsted becomes manager of Rancho Las Mariposas, a mining estate in the Sierra Nevada mountains of California; visits what becomes Yosemite, later writes a report that inspires interest in Yosemite becoming the first national park. |
| 1865 | General Robert E. Lee surrenders at Appomattox, Virginia; conclusion of Civil War (April). |
| 1865 | Landscape design firm of Vaux & Olmsted is formed; begins to work on parks and other projects. |
| 1865 | Affleck begins working to rebuild economic and agricultural conditions in Texas, which was devastated by the Civil War. |
| 1865–1868 | Affleck travels to Europe to promote an immigration scheme among English and Scots to replace African-American agricultural workers in the American South with immigrants from Great Britain. |
| 1868 | Affleck seeks European backing for a meat-packing scheme related to his immigration scheme; establishes an office in Galveston to implement scheme. |
| 1868 | Affleck completes *Hedging and Hedging Plants.* |
| 1868 | Affleck dies on December 30 at Glenblythe. |

| 1869 | *Hedging and Hedging Plants, in the Southern States* appears, intended to be the first in a series of works for the southern gardener. |
|------|-----|
| 1883 | Olmsted moves his office to Brookline, Massachusetts; continues to work throughout the U.S. |
| 1893 | Olmsted becomes design team leader for the 1893 World's Columbian Exposition in Chicago, his last project. |
| 1895 | Olmsted retires from active practice. |
| 1903 | Olmsted dies in the sanatorium for which he had designed the grounds. |

A NOTE ON THE
# THOMAS AFFLECK PAPERS

Conventionally, archivists speak of archival collections as sources of evidence and information. Scholars come to these primary sources for a record of past events and to interpret those events, using their findings to develop their arguments and provide supporting evidence. Alternately, and more organically, scholars may find inspiration among family or personal papers that unexpectedly sets them on a course of inquiry when a heretofore-unexamined topic or event presents itself among the pages of a diary or correspondence. Such was author Lake Douglas's experience with the Thomas Affleck Papers, held in the Louisiana and Lower Mississippi Valley Collections (LLMVC) of the LSU Libraries' Special Collections.

The collection, which the Libraries have owned since 1934, is comprised of Affleck's incoming and outgoing correspondence, legal documents, diaries, record books, writings, and selected publications. Thomas Affleck emerges from these thousands of items as an inveterate experimenter, creative inventor, entrepreneurial and hard-nosed businessman, meticulous nurseryman and horticulturist, influential advocate for scientific agriculture, and prolific writer. It is this last, Affleck's output as a writer, that is the focus of this book.

Examination of Affleck's archive reveals his broad interests and wide-ranging web of contacts. It also illustrates his thinking about southern agriculture and horticulture and the importance of developing them, his scientific and logical approach to both agriculture and business, and his abiding interest in sharing what he discovered through his own experimentation and learned via his associations. These characteristics suited him well to write on a variety of topics of interest to an audience of farmers, gardeners, plantation owners, and horticulturists.

Indeed it is clear from Affleck's correspondence that he considered his work a science, and he sought to spread his gospel of scientific agriculture through his many writings and publications. As Douglas's compilation of

Affleck's works shows, he was a frequent contributor to and editor of agricultural journals and columns. He is also noted for authoring *Affleck's Southern Rural Almanac and Plantation and Garden Calendar*, which began publication in 1851. The almanac detailed planting instructions for plantation cash crops, fodder crops, the kitchen garden, and ornamental gardens and included planting dates for both Natchez, Mississippi, and New Orleans, Louisiana. It enjoyed a wide circulation through 1861. In addition to these contributions to the agricultural literature, Affleck extended his influence with the publication of specialized record books for documenting plantation work. Describing himself as "having been trained in Scotland to the strictest business habits," he found unconscionable the lack of standardized record keeping among both his own plantation overseers and in general plantation practice. This frustration led him to create his *Affleck Plantation Record and Account Book*, which first appeared in 1847. The books proved popular with planters, and thanks to their wide adoption, they show up in plantation papers held in archives across the South. Scholars have found them to be an excellent source for researching antebellum southern agriculture, slavery, and plantation economy and operation.

In his introduction, Lake Douglas argues that while Affleck's personal correspondence and business archive held by Special Collections provide the details of his life and career, it is Affleck's writings scattered in rare and obscure publications that yield the best evidence of his place in nineteenth-century American life. Douglas's observation nicely encapsulates the idea behind Special Collections' Louisiana and Lower Mississippi Valley Collections. As an integrated research collection, the LLMVC combines regionally focused, comprehensive published holdings and extensive archival collections. These materials complement and support each other, and together they foster in-depth study and analysis of multiple facets of our collective past.

Tara Zachary Laver, MLIS, CA
Curator of Manuscripts
Special Collections, LSU Libraries

# INDEX

Abbé Maillard rose, 190

Abricoté rose, 192

*Acacia dealbata,* 211n

acclimation, 134, 152–55

accounting, agricultural, 133

*Achimenes,* 166n

Adam rose, 192

Affleck, Isaac Dunbar (son), 2–3, 106, 120, 219

Affleck, Mary Hannay (mother), 245

Affleck, Thomas

—life and career: birth and youth, xiii, 5–6, 245; broad interests, xvii, xxii, xxiv–xxv, 9, 251; business and commercial dealings, xv–xvi, 7, 16–17, 18, 22–23, 34, 111; chronology of, 245–49; as Civil War civilian leader, 25, 28; Cole article on, 19–30; correspondence, xvi, 251; death, xiii, xviii, 27, 248; destruction of nursery stock, 237–38, 244, 248; Downing and Olmsted compared to, xvii–xx; financial problems, 7, 22–23, 25; influence of, 12, 18, 29, 104, 111, 238, 252; library, 8–9; marriage, 7, 246; move to Cincinnati, xiv, 245; move to Galveston, 26–27, 248; move to Mississippi, 104, 246; move to Texas, 23, 27–28, 237–38, 244, 247; national reputation, 6, 11–12, 24; as nurseryman, xv–xvi, 7, 11, 27, 251; personal faults, 29; as prolific writer, 20–21, 251; salary, 16; scholarly attention to, xx–xxi, xxii, 1–3; as slaveowner, 25; Texas plantation of, 25–26; trip to South, 78–83, 246; as visionary, xiii, 2, 30; Williams article on, xxii, 5–18; writing style, 38, 237

—works and writings, xvi–xvii, 251–52; for

*American Agriculturist,* 109–12, 113–25; *Bee-breeding in the West,* 52, 58, 70, 246; for Dallas *Herald,* 231–34; *Hedging and Hedging Plants, in the Southern States,* 132n, 221, 235–44, 248; for Houston *Telegraph,* 219, 222–31; for New Orleans *Commercial Times,* 16; for New Orleans *Daily Picayune,* 157, 159–67, 172–73; for New Orleans *Weekly Picayune,* 8, 9–10, 16, 21; *Plantation Record and Account Book,* 13–14, 21–22, 24, 133, 157, 246; *Southern Rural Almanac,* xvi, 12–13, 16, 21–22, 23, 132, 172, 173, 198, 243, 247, 248; uncompleted manuscripts, 15–16; for *Western Farmer and Gardener,* xiv–xv, 5–6, 34, 39–107, 108

*Affleck's Southern Rural Almanac and Plantation and Garden Calendar,* 16, 132, 172, 173, 198, 243, 247, 248; circulation of, 21–22; contents of, xvi, 12–13; finances of, 23

Agricultural, Horticultural, and Botanical Society of Jefferson College, 80–81, 88–89, 103, 105–7

agricultural costs, 131, 136, 151, 226, 229–30

agricultural fairs, 80, 85–88, 103–4, 105–7

*Agricultural History,* 2

Agricultural History Society, 2

agricultural implements, 28, 87, 111, 116–19; northern-made, 117–18

agricultural newspapers and magazines, 19, 33–37, 41, 58

agricultural prices, 138–39, 143. *See also* land prices

agricultural reform and improvement movement, 8, 126, 129–30, 131

agricultural societies, 2, 39–40, 80–81, 127